Soft Computing in Wireless Sensor Networks

Soft Computing in Wireless Sensor Networks

Edited by

Huynh Thi Thanh Binh

Nilanjan Dey

CRC Press
Taylor & Francis Group
Boca Raton London New York

CRC Press is an imprint of the
Taylor & Francis Group, an informa business
A CHAPMAN & HALL BOOK

CRC Press
Taylor & Francis Group
6000 Broken Sound Parkway NW, Suite 300
Boca Raton, FL 33487-2742

Printed on acid-free paper

International Standard Book Number-13: 978-0-8153-9530-0 (Hardback)

Library of Congress Cataloging-in-Publication Data

Names: Thanh Binh, Huynh Thi, editor. | Dey, Nilanjan, 1984- editor.
Title: Soft computing in wireless sensor networks / editors, Huynh Thi Thanh Binh, Nilanjan Dey.
Description: Boca Raton : Taylor & Francis, a CRC title, part of the Taylor & Francis imprint, a member of the Taylor & Francis Group, the academic division of T&F Informa, plc, 2018. | Includes bibliographical references and index.
Identifiers: LCCN 2018046892| ISBN 9780815395300 (hardback : acid-free paper) | ISBN 9780429438639 (ebook)
Subjects: LCSH: Wireless sensor networks. | Soft computing. | Internet of things.
Classification: LCC TK7872.D48 .S637 2018 | DDC 006.2/5--dc23
LC record available at https://lccn.loc.gov/2018046892

Visit the Taylor & Francis Web site at
http://www.taylorandfrancis.com

and the CRC Press Web site at
http://www.crcpress.com

Contents

Preface

Recently, the Internet of Things (IoT) has received huge interest from scientists and technology designers throughout the world and is considered the next network generation in the near future. This concept starts from the idea of building a network consisting of all real-life objects (things) in which they can connect, interact, incorporate, and communicate data and information, accomplishing complex tasks. This way, the strength of the Internet is fully investigated in a highly effective way, bringing in practical applications in many fields such as economy, information technology, military, etc. In practice, some specific applications of the IoT, namely smart home, smart traffic, etc., have proven its feasibility, convenience and promising benefits.

However, the IoT can only be successfully applied within the support of wireless sensor networks (WSNs). Today, the development of the Internet, radio communications, and information technology has brought the potential development of WSNs. The sensor network might include different types of sensors, such as earthquake sensors, magnetic field sensors, low-rate sampling sensors, camera sensors, infrared sensors, sound sensors, radar, and so on, which can monitor large areas under different conditions.

This volume is comprised of nine chapters, providing different techniques for solving optimization problems in WSNs. An overview about WSNs is given in Chapter 1, followed by an introduction to optimisation problems in WSNs in Chapter 2, including major factors that need optimisation; problem classification and well-known approaches to solve them; and current achievements and results. These first chapters provide readers with basic theoretical knowledge about the topic and why it is demanding to push the boundaries in this field. Chapter 3 discusses the application of machine learning (ML) algorithms in WSNs, which play a crucial role in WSNs as they simplify huge amounts of data generated and gathered at the central nodes in WSNs. To elaborate the importance of using soft computing in this research topic, Chapter 4 presents the relevance of soft computing techniques in the significant management of WSNs and introduces soft computing techniques like ant colony optimization and particle swarm optimization (PSO), artificial neural networks (ANNs), perception, fuzzy logic, evolutionary computing, and genetic algorithms in the management of WSNs. There is also a broad analysis of this intelligence-based soft computing strategy as connected in the different operational parts of wireless sensor networks. After that, the following chapters discuss more narrowed optimization problems in WSNs including their challenges, solutions, and achieved results. Chapter 5 concerns intrusion detection systems in mobile ad hoc networks, in which the definition of mobile ad hoc networks and different kinds of attacks are well-elaborated. The performance of grammatical evolution is also analysed with ad hoc flooding and route disruption attacks on various mobility patterns of the nodes on the network. Chapter 6 presents an introduction to coverage optimisation in WSNs. The problem is classified into smaller categories of area coverage, target coverage and barrier coverage including an in-depth discussion about state-of-the-art solutions and recent achievements. Chapters 7 and 8 concern cluster formation and sensor positioning in WSNs, respectively. For any cluster-based routing technique, the major challenge is to efficiently elect the cluster head (CH) nodes. The wrong choice of CHs lead to early death of nodes and network collapse. Chapter 7 addresses this problem and introduces an energy efficient protocol for cluster formation in WSNs. Chapter 8 discusses alternative effective and efficient techniques to replace the traditional ones that are more expensive, more energy-consuming, face connectivity failure problems, and are less accurate in performance. This chapter mainly focuses on sensor deployment strategies to achieve optimum coverage areas and minimize energy consumption. Finally, Chapter 9 is concerned with an elaborate and illustrative discussion about the application of IoT in the healthcare sector. The IoT healthcare system acts as a platform between wearable and implantable body sensor networks (WIBSNs) with specific needs. This chapter examines the various attributes associated with WIBSNs and analyses its workings with respect to healthcare. It also gives the reader an overall perspective of the entire ecosystem of IoT–based healthcare and the functioning of the associated components.

We would like to express gratitude to the authors for their contributions. It would not have been possible to reach this publication quality without the contributions of the many anonymous referees involved in the revision and acceptance process of the submitted manuscripts. Our gratitude is extended to them as well. As the editors, we hope this book will stimulate further research in medical imaging applications-based algorithmic and computer-based approaches and utilize them in real-world clinical settings. We would also like to thank the reviewers for their diligence in reviewing the chapters. Special thanks go to our publisher, CRC Press/Taylor & Francis Group.

We hope that this book will present promising ideas and outstanding research results supporting further development of soft computing approaches in solving optimization problems in WSNs.

Huynh Thi Thanh Binh, PhD
Hanoi University of Science and Technology, Vietnam

Nilanjan Dey, PhD
Techno India College of Technology, Kolkata, India

Editors

Huynh Thi Thanh Binh is an associate professor and vice dean of the School of Information and Communication Technology (SoICT) at the Hanoi University of Science and Technology (HUST). She is head of the Modeling, Simulation and Optimization Lab. Her current research interests include computational intelligence, artificial intelligence, memetic computing and evolutionary multitasking. She has published more than 80 refereed academic papers/articles and 2 books and is the editor of 1 book. She is associate editor of the *International Journal of Advances in Intelligent Informatics, VNU Journal of Science: Computer Science Communication Engineering* and serves on the Editor Board of the *Journal of Computer Science and Cybernetics*. She has served as a regular reviewer, and program committee member of numerous prestigious academic journals and conferences, such as *IEEE Transactions on Vehicular Technology and Journal of Information Science and Engineering*. She is chair of IEEE Computational Intelligence Society Vietnam Chapter and a member of the IEEE Women in Computational Intelligence subcommittee (2017–2018). She is a member of a few IEEE Asia Pacific committees: Strategic Planning, Membership Development, Individual Benefits and Services, and Humanitarian Technology Activities.

Nilanjan Dey was born in Kolkata, India, in 1984. He received his bachelor's degree in information technology from West Bengal University of Technology in 2005; his master's degree in information technology in 2011 from the same University, and his PhD in digital image processing in 2015 from Jadavpur University, India.

In 2011, he was appointed as an assistant professor in the Department of Information Technology at JIS College of Engineering, Kalyani, India, followed by the Bengal College of Engineering and Technology, Durgapur, India, in 2014. He is now employed as an assistant professor in the Department of Information Technology at Techno India College of Technology, India. His research interests are signal processing, machine learning, and information security.

Dr. Dey is an Associate Editor of *IEEE Access* and is currently the editor in-chief of the *International Journal of Ambient Computing and Intelligence, International Journal of Rough Sets and Data Analysis* and the *International Journal of Natural Computing Research*. He is the Series Co-Editor of *Springer Tracts in Nature-Inspired Computing* (STNIC), Springer; Series Co-Editor of *Advances in Ubiquitous Sensing Applications for Healthcare* (AUSAH), Elsevier, and Series Editor of *Computational Intelligence in Engineering Problem Solving* and *Intelligent Signal Processing and Data Analysis*, CRC Press (FOCUS/Brief Series), and *Advances in Geospatial Technologies* (AGT) Book Series, (IGI Global), US.

Contributors

B. Anuradha
Department of Electronics and Communication Engineering
SV University College of Engineering
Tirupati, India

Balamurugan Balusamy
School of Computing Science and Engineering
Galgotias University
Greater Noida, India

V. Bapuji
Department of Computer Science
Vaageswari College of Engineering
Karimnagar, India

Huynh Thi Thanh Binh
School of Information and Communication Technology (SoICT)
Hanoi University of Science and Technology
Hanoi, Vietnam

Anupkumar M. Bongale
D Y Patil College of Engineering
Ambi, India

G. Bhanu Chander
Department of Computer Science and Engineering
Pondicherry University
Pondicherry, India

Nilanjan Dey
Department of Information Technology
Techno India College of Technology
Kolkata, India

D. Praveen Kumar
Computer Science Department
Indian Institute of Technology (ISM)
Dhanbad, India

G. Kumaravelan
Department of Computer Science and Engineering
Pondicherry University
Pondicherry, India

Ramchandra S. Mangrulkar
Dwarkadas J. Sanghvi College of Engineering
Mumbai, India

B. Manjula
Department of Computer Science
Kakatiya University
Warangal, India

Nguyen Hai Nam
Uppsala University
Uppsala, Sweden

Prashil D. Negandhi
Dwarkadas J. Sanghvi College of Engineering
Mumbai, India

Nirmala C. R.
Bapuji Institute of Engineering and Technology
Davangere, India

Bibudhendu Pati
Department of Computer Science and Information Technology
Siksha 'O' Anusandhan University
Bhubaneswar, India

Binod Kumar Pattanayak
Department of Computer Science and Engineering
Siksha 'O' Anusandhan University
Bhubaneswar, India

T. Poongodi
School of Computing Science and Engineering
Galgotias University
Greater Noida, India

N. Pushpalatha
Annamacharya Institute of Technology and Sciences
Tirupati, India

K. Ramani
Department of I.T.
Sree Vidyanikethan Engineering College
Tirupati, India

Mamata Rath
C. V. Raman College of Engineering
Bhubaneswar, India

Anu Rathee
School of Computing Science and Engineering
Galgotias University
Greater Noida, India

D. Srinivas Reddy
Department of Computer Science
Vaageswari College of Engineering
Karimnagar, India

D. K. Sah
Computer Science Department
Indian Institute of Technology (ISM)
Dhanbad, India

Chaya Shivalingagowda
Department of Electronics and Communication
GITAM University
Visakhapatnam, India

Monika Yadav
School of Computing Science and Engineering
Galgotias University
Greater Noida, India

1

Introduction to Wireless Sensor Networks

G. Bhanu Chander and G. Kumaravelan

CONTENTS

1.1 Introduction

As the result of significant developments in the field of electronic communication, computer science, and information technology, we now have a new computing and communication architecture known as wireless sensor networks (WSNs). In the past, wired sensors implemented into restricted appliance areas produced low results only. Meanwhile, the convention of adapting wireless technology produced sensor nodes which formulated more feasible solutions prior to wired sensor nodes. Recently, there is a measurable benefit with an immense possibility of connecting various devices and networks together in an eloquent effort to address important issues and challenges.

The first wireless sensor network was a sound surveillance system developed by the US military in the 1950s to detect and track Soviet submarines [1]. This servicing technology is still at work today for monitoring undersea wildlife and volcanic activity. Nowadays, with the rapid development of the aforementioned WSN, this technology is easily deployed in the largest geographical area that allows alike sensor nodes to gather the raw data and communicate and transfer it in new applications [2].

1.2 Wireless Sensor Networks

Wireless sensor networks are a collection of undersized individual sensor nodes. Depending on the application scenario, this number may be raised to hundreds to thousands apiece, where a node is connected with additional nodes. Each sensor node in a wireless sensor network observes its environmental phenomena and this collected information is transmitted to one or more sink stations through a wireless link depending on the network deployment.

There are three main functionalities of sensor node: the first is sensing its neighboring atmosphere, the second is processing the data which was initially observed, and the third is communicating or connecting through

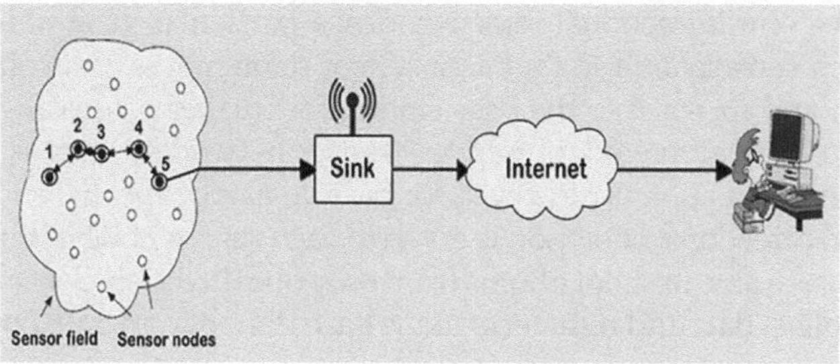

FIGURE 1.1
Communication architecture of a wireless sensor network.

additional sensor nodes or sink stations within the network. Among these above specifications, sensing the environment is the most important one.

WSNs can be structured in centralized, distributed, and ad hoc ways. Figure 1.1 shows the general communication structure of a WSN. It primarily consists of sensor field, sink node, and user or task management component. Sensor nodes are deployed in a specific area is known as a sensor field. Each sensor node in the sensor field is capable of sensing its own respective environment phenomenon and routing the observed data to the sink node through a multi-hop connection. A sink node is a special kind of sensor node which collects data from the sensor field and performs the necessary operations or simply forwards the collected data to the user or task manager node. In addition, it also sends queries to the sensor nodes in the network and retrieves the required data. Finally, the user or task manager node accords an assignment to each sensor node in the sensor field. Nevertheless, the sink node connects with the user or task manager node via an Internet or satellite connection [1,2].

1.3 Types of Sensor Nodes

The sensor is used in a large range of constraints which correspond to healthiness forms, such movements, electrical signals, thermal, and compelling energy.

Sensor nodes are mainly two types (1) passive sensors and (2) active sensors.

1.3.1 Passive Sensors

Passive sensors observe the data by active probing, which means their energy is used only to amplify the analog signals. Passive sensors don't require external resources as their energy is used only to generate analog signals. These sensors can change their physical properties (like capacitance) and generate electrical signals. Again, these passive sensors are divided into two types' *passive omnidirectional sensors* and *passive narrow beam sensors*.

For passive omnidirectional sensor nodes, a particular geographical area is given to each sensor node so that area sensor node moves and collect reliable information and forward to the base station for further processing. The present theoretical research work in wireless sensor networks assumes that sensor nodes are omnidirectional. However, for passive narrow-beam sensor nodes, a prearranged direction or notation is given to each sensor node in the same way that cameras are able to shoot photos from only one direction. Sensor nodes can move and collect data and transform to the base station from only one direction.

1.3.2 Active Sensors

Active sensor nodes sense the data by manipulating the environment which means they need continuous energy to generate analog signals. Active sensor nodes dynamically probe the neighboring environment and use an external power supply of energy, which is needed to monitor and operate. Digital sensors generate distinct indications which are later converted to digital characterization based on the parameters measured. A single byte group discrete values output is measured as the quantity. Analog sensors generate continuous signals (such as temperature, pressure, and water flow), which are measured as continuous analog signals.

Each sensor node is also called a *mote* but all motes are not always called sensor nodes. Motes receive and transmit the data and employs some processing techniques inside the sensor node, like compression, encryption etc. So, sensor node operate like motes, but motes do not always operate as sensor nodes [1–3].

1.4 Sensor Node Description

A sensor node is a tiny device that consists of four major hardware components, namely the sensing unit, processing unit, transceiver, and power generator. In addition, the sensor node also uses location devices such as global positioning system (GPS) and a mobilizer. Figure 1.2 shows the basic configuration of a sensor node.

1.4.1 Sensing Unit

A sensing unit observes its environmental phenomena for gathering information (such as temperature, pressure, light, and displacement) as well as producing the corresponding output as an optical or electrical signal. A sensing unit basically consists of two parts; a sensor and an analog to digital converter (ADC). In the sensor part, one or more sensors there produce the analog signals, and these analog signals are transferred into digital signals

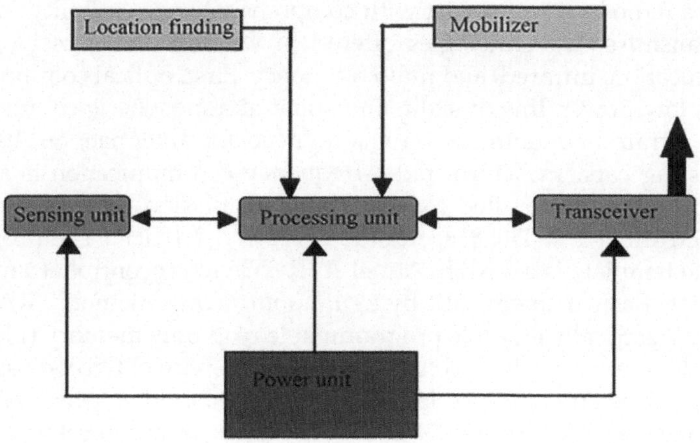

FIGURE 1.2
Basic configuration of sensor node.

through the ADC and fed into a processing unit for supplementary operations. Based on the observed or gathered phenomena, sensor nodes are classified as thermal, optical, acoustic, or mechanical sensors.

1.4.2 Processing Unit

A processing unit includes two parts; storage plus a processor. A storage component, which works as nonvolatile memory used to store programming instructions, temporarily stores sensed data. Also, it will store processed data on occasion. The processor in sensor node makes the sensor node collaborate with other nodes in the network to complete the designated exercise. The microcontroller performs tasks, processes the data, and controls the other functionalities of a sensor node. Because of some special characteristics of microcontrollers, like low cost, easy attachment with supplementary apparatuses, effortlessness of training, and squat power utilization, microcontrollers are used in sensor nodes. ATMEL, At mega 128L, and MSP430 are the most commonly used microcontrollers because they have power saving capabilities. Among these, MSP 430 has six different power modes from wholly energetic to wholly powered down. These kinds of power reduction approaches enlarge the system lifetime and make the sensor node live longer.

1.4.3 Transceiver Unit

The transceiver attaches the node to the network. It is an amalgamation of transmitter and receiver to a solitary device which works on specific radio frequency (RF). The RF message requires modulation, filtering, multiplexing, band bass, and demodulation, which makes it more complex and expensive. Basically, sensor node uses industrial scientific medical bands (ISM) which provide a radio band at no cost with comprehensive accessibility. There are three transmission technologies offered in wireless transmission; optical communication, infrared, and radio frequency. First, optical communication requires low power, line of sight, and squat at atmospheric circumstances. Second, infrared communication has no need for antennae, but has a low broadcasting capacity. Third, radio frequency communication is the most relevant method best suited for wireless transmission because it provides free spectrum at 173, 433, 868, and 915 MHz and 2.4 GHz. The progression of a mote template is a 4 MHz Atmel AVR 8535 microcontroller amid 8 KB instruction flash memory, 512 bytes random access memory (RAM) and 512 bytes electrically erasable programmable read only memory (EEPROM). Tiny OS Operating systems consist of 3500 byte space of OS code as well as 4500 byte space offered for code. The evolution of another sensor node template, namely an AMPS posse 59–206 MHz SA-1110 microprocessor, is run on a multithreaded operating system [2–4].

The majority of sensor nodes use low-rate, wireless personal area networks which are IEEE 802.15.4 standard. Standard meaning the transceiver has four operational modes, like transmit, receive, idle, and sleep. In receive and idle mode, the power consumption of the transceiver is the same. However, the transceiver should initiate sleep mode and not be left in idle mode when it is not actively working, otherwise considerable energy will be wasted. It will simply interchange from sleep state to the active broadcast or receive state [1].

1.4.4 Power Unit

The sensor node is a small microelectrical appliance which consists of low energy. The power unit sustains the sensor node in a rough and unconditional environment where changing batteries is expensive and difficult. Mainly, sensor nodes use their energy for sensing, transmission, and data aggregation purposes. Among these aforemention operations, the transmission of information consumes the majority of energy, compared to the other operations. The power source of a sensor node is constituted by a rechargeable or nonrechargeable battery. Rechargeable sensor nodes are capable to renovate their power from the nature circumstances like solar, high temperature differences, and pulsation. Non-rechargeable sensor nodes must be competent to operate until either the assignment time is completed or the battery is replaced. However, based on the application scenario, the lifetime of sensor node will be decided. For example, scientists monitoring the presence or age of ice (especially in the form of glaciers) or observing the bed slides in the ocean need the sensors able to work continuously. When the sensors are deployed in the battlefield, finding the temperature in particular area may need a few hours or days. In the past, the batteries of tiny sensor nodes were manufactured with vanadium and molybdenum oxide. Future energy exploration from the environment occured because sensor nodes operated in irregular places. The battery for sensor nodes should be as small and efficient as possible. Nickel–zinc, lithium–ion, and nickel–metal hydride electrochemical objects are used for their electrodes.

In addition, some routing protocols and specific applications need the location of a sensor node because they are generally deployed in unconditional areas and need to collaborate with other sensor nodes about their present location for the transmission of data, which is possible by a location finding system. A mobilizer is an optional component in the description of a sensor node which moves the sensor node from one place to another place to complete the assigned task.

1.5 Applications of Sensor Nodes

Wireless sensor networks are deployed in an irregular environment with a wide range of applications. There are many types of sensor nodes (seismic, acoustic and thermal, radar, visual, etc.) in WSNs which extend their basic applications. WSNs are comprehensively used in areas such as environment monitoring, tracking, crisis management, monitoring patients, radioactivity detecting, detecting dangerous material such as explosives, security and surveillance, underground material structures, critical infrastructure, office and home automation, traffic control, disaster and flood intimation, and identification [5–7].

1.5.1 Military Applications

Wireless sensor network implementation primarily began with the military application with the formation of DARPA and enemy tracking. Since sensor nodes are densely deployed, if some sensor nodes are destroyed by antagonistic actions, it doesn't influence or affect the military operation. This makes sensor nodes ideal for the battlefield. Using appropriate sensors in the specific network area, the identification of enemy forces, movement detection, analysis, and progress of their movement can be achieved. Sensor nodes provide the following services in the battlefield [5]:

Battlefield Surveillance: Information from the border, battlefield and some other vital behavior in the area will be gathered by the sensor node.

Targeting: Sensors are placed in weapons and collect information about targets (like motion, movement, angle, distance, etc.) which can be sent to the shooter. This sensor is used for better target estimation.

Intrusion Detection: Sensor nodes are used to identify the intrusion in the network and sounding an alarm. The defense system will then assess an appropriate recourse for the attack.

Monitoring Forces: Sensor nodes monitor the movement of troops and tanks, and report that information to the base station.

Target Classification: Some particularly designed sensors can monitor suspicious objects to better track potential enemies. An endocrine-based intelligent distributed cooperative algorithm (EIDCA) is proposed for soldier tracking.

Battle Damage Evolution: Sophisticated sensor nodes can be deployed to estimate the damage of battle affected areas.

1.5.2 Environmental Observation

Sensor nodes play an important role to supervise and estimate the environmental conditions of large geographic areas (like temperature, humidity, rainfall, etc.) which help in pollution monitoring, forest fire detection, and flood detection. The first well-known wireless sensor network for weather coverage was an automated real-world evaluation in real time (ALERT). ALERT can produce real-time information concerning rainfall and water levels to estimate the possibility of flooding. Environmental monitoring is broadly divided into two types; indoor and outdoor monitoring. Indoor monitoring is used for buildings and offices. This application involves sensing humidity, temperature, and air quality. Examples of outdoor monitoring applications are volcanic eruptions, earthquakes, chemical hazard detection, and weather forecasting [6].

1.5.3 Forest Fire Detection

Sensor nodes placed in forests are used to detect fires. Some specially designed sensor nodes are used to measure the temperature, humidity, air direction, and the speed of air, which helps to determine the level of fire risk. Wireless sensor networks locate the starting point of the fire and alert the respective authorities to put it out.

1.5.4 Pollution Monitoring

Wireless sensor networks can be employed towards monitoring and examining the pollution intensity in a particular area. Sensor nodes, which are deployed around the city, can be used to estimate the respective environmental factors, and accordingly producing a warning signal when these factors reach their aforementioned threshold limit. For example, LTE-M module–based Zigbee wireless sensor nodes are kept on the stations of public vehicles to collect data, and are sent to the cloud base station for further processing of the pollution level.

1.5.5 Industrial Monitoring

Industries are mostly using wireless sensor networks for determining the level of output quality in its operation and also in cost-reduction procedures. For example, in nuclear power projects, sensors are used to monitor the water level in the tank and the temperature and pressure in refrigerators. One more central application of sensor networks is to supervise machine health. The aspiration of this application is to detect the defective parts of equipment which need to be repaired or replaced. Inventory control is another important problem in big industries. The globalization of larger industries makes it difficult to manage their equipment and products, so, the management of these companies achieved through wireless sensor networks.

1.5.6 Agriculture Monitoring

The use of wireless sensor networks in the agriculture research community has been increasing rapidly. Sensor nodes sense parameters like temperature, humidity, soil moisture, and sunlight of the agriculture field so the necessary precautionary measures will be given to its stakeholders to increase the harvest. The LOFAR-Agro project is used in potato fields for crop monitoring. Here, wireless sensor networks are used to detect where the potatoes are infected by disease. By locating such areas, the use of pesticides is minimized and confined to susceptible areas.

1.5.7 Health Monitoring

In this application, patients are outfitted by collective sensors on special points of their anatomy to supervise patient metrics, such as blood pressure, heart rate, etc. Home-based wireless ECG monitoring systems exploiting Zigbee technology are considered some of the applications that take advantage of health monitoring. Such methods can helpful for supervising people in their homes, along with periodic supervising by a general practitioner. Electrocardiograms (ECG) are imperative for detecting heart disease abnormalities. Clinical ECG equipment may be functional for short-range inspection, but are impractical for home health use. In recent times, wireless ECG supervising systems have been created that are compatible with Bluetooth, as well as Zigbee, protocols. A Zigbee device is capable to advocate unfailing multi-hop, self-organizing, and mesh network which steadies the appliance software layers considered by the IEEE 802.15 plus Zigbee alliance. Zigbee can provide with diminutive energy connectivity along with an undersized rate for the equipment that necessitates prolonged battery continuation [8].

Wireless capsule endoscopy (WCE) suggests an illustrated inspection of the entire gastrointestinal (GI) tract and the rationalized analytical progression of undersized bowel illness, moreover the triumph over the boundaries of conventional diagnosing equipment. In order to strengthen the WCE, computational procedures can be implemented for specific capsule localization and tracking. For perfect understanding of the wireless capsule endoscope (WCE) location point toward the wrongdoing position which plays an essential role for dissimilar reasons including perfect localization of injuries plus mechanized CE direction-finding. Localization is fulfilled by several localization techniques such as received signal strength (RSS)–based localization, time of arrived (TOA)–based localization, magnetic-based localization, image-based localization, and date of arrived (DOA)–based localization [9].

1.5.8 Smart Home

Remote management of home applications is achievable by using some integrated, specially designed sensor nodes. Sensor nodes can be deployed in household devices like microwaves and washing machines, and can be maintained without the need for human interaction. Sensor nodes share some common resources like water, heaters, and electricity in the intelligent home projects.

1.5.9 Power Grids

Wireless sensor nodes deployed in power grid applications provide necessary information regarding eclectic utilization, which makes them more efficient. Sensor nodes are used for measurement and tracking energy production and consumption to monitor the entire surrounding area. If there are any abnormal positions detected, the nodes will send that diagnostic information to an authority to solve these problems.

1.5.10 Automobiles

Sensor networks have been used to monitor and track vehicles for a long time. A common example would be video cameras that are fixed at a particular area to observe traffic. The videos are sending to a human operator, and as a low-cost replacement for video cameras, small sensor nodes are installed to monitor traffic. Sensor nodes take pictures of traffic, which are sent to a human operator or automatic controllers, and take control signals. Sensor nodes are installed in particular vehicles to monitor and track their movements.

1.6 Protocol Stack of WSNs

In WSNs, sensor node energy consumption has an effect on the configuration of protocol layers in addition to managing the sensed data at every level. Protocol stacks (see Figure 1.3) are used in sensor and sink nodes inside the network and consist of a physical layer, data link layer, network layer, transport layer, application layer, power management plane, mobility management plane, and task management plane [1–4].

1.6.1 Physical Layer

The physical layer is designed for modulation, carrier frequency generation, and transmission, as well as for receiving techniques, signal detection, and frequency selection. The transmission links of wireless sensor networks are twisted based on infrared, optical, and radio. Sensor nodes present employment based on radio frequency design. So, the physical layer typically takes care of communication-based radio links.

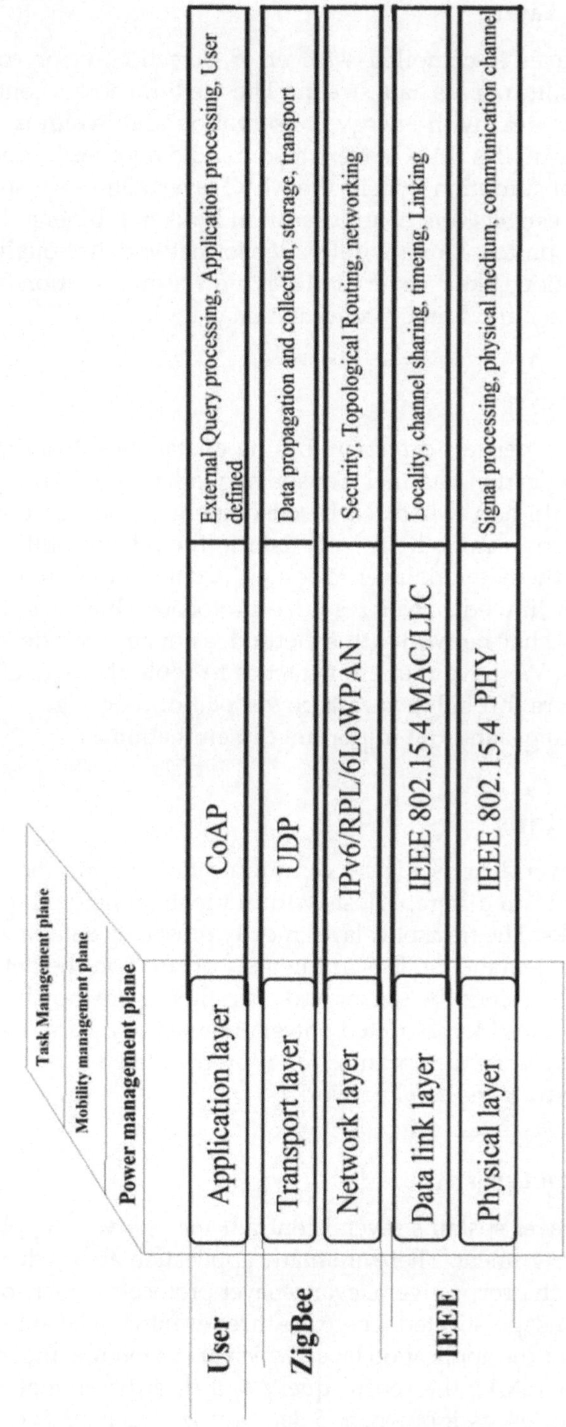

FIGURE 1.3
WSN protocol stack suite.

1.6.2 Data Link Layer

The data link layer is compiled with error detection, error control, and medium-access control plus data stream. The medium access control (MAC) layer set of rules deal with energy conservation and avoiding a collision with neighboring nodes. MAC layers also handle another important issue called energy consumption, where the MAC layer allows sensor nodes to self-organize by establishing communication links hop-by-hop. For designing a MAC layer protocol for a wireless sensor network, we ought to obtain dissimilar limitations like recovery failures, power preservation (suitable for low-power devices), and limited computational ability.

1.6.3 Network Layer

The network layer permits direction-finding of statistics data throughout a wireless communication channel. Sensor networks are without infrastructure and are multi-hop, where each sensor node as source node collects sensed data and transmits the sensed data to the reliable sink node. So, it is important for the network layer to route the data in the direction of sink node. Because of limited power resource protocols which are designed in traditional and ad hoc networks, this method is not appropriate for wireless sensor networks. We have numerous tactics to route the data as a routing power expenditure through a vacant power platform on vigor metrics, and data-centric routing stands on important dissemination.

1.6.4 Transport Layer

The transport layer affords logical communication among the application process operations on different hosts within the main network and supplementary networks. The transport layer mostly preserves data flow when the application layer demands it. Designing a set of rules on the network layer is very complicated, since the sensor node parameters are mostly affected by numerous limitations like restricted power and memory. This layer is principally compulsory, while the organization is premeditated to be contacted via the Internet or extra peripheral networks.

1.6.5 Application Layer

The application layer sustains several contradictory software applications on the sensibility assignment. There are many application areas which are well-designed, although prospective relevance layer protocols for sensor networks are still laregly unsophisticated. There are three eventual relevant layer procedures designed for the application layer, which are as follows: the sensor management protocol (SMP), the sensor query and data dissemination protocol (SQDDP), and the task assignment and data advertisement protocol (TADAP).

1.6.6 Power Management Plane

The power management plane supervises how sensors allocate energy resources, moreover they assign energy resources to sensor nodes to complete their individual assignments. For instance, sensor nodes may turn off their receivers after acquiring a message from one of their bordering nodes. This procedure prevents duplicate messages.

1.6.7 Mobility Management Plane

The mobility management plane discovers and stores each node's progress as well as assigns sensor nodes in the sensor field. Therefore, a backward route to the user is constantly preserved and sensor nodes are able to track their bordering sensor nodes.

1.6.8 Task Management Plane

A task management plane assigns a task for each sensor depending upon its schedule and balances each task completed in the assigned time. The task management plane connects to the sink node, base station, or user via Internet or satellite.

1.7 Security Requirements in WSN

The following are some most important requirements in WSNs [4–6,10–15]:

Data Confidentiality: Data confidentiality protects the data so any unauthorized user can't read and investigate it. Attackers should not know the frequency and content of facts broadcasted. Moreover, the data should acknowledge its anticipated receiver only; sensor identifiers like public key distribution should be done in a secure way. Additionally, secure channel generation and a strong key distribution algorithms are essential. Without permission, any nearby nodes should not be able to read the content of other nodes in the network.

Data Integrity: Data integrity guarantees that data are not personalized by any unapproved third party, meaning that data throughout the transmission from the source to destination should not be modified or misused by an attacker. Malicious nodes may change substantial messages so that the redesigned data will not send to an authentic source. Therefore, there is need to verify the data occasionally.

Self Organization: Sensor nodes should be capable of self-organization because they can't be monitored and maintained after deployment. Wireless sensor networks are a type of ad hoc network; therefore, applying traditional cryptographic algorithms is difficult. Moreover, if sensor nodes are behaving enthusiastically, sharing keys among nodes becomes complex ahead of deployment if a symmetric key is used.

Data Freshness: Sensor nodes should make sure that messages which are transmitted or received do not receive old messages. Suppose that in symmetric key cryptography, occasionally previous data is sent repetitively; if a shared key is used, it is compulsory to upgrade the data over time. To triumph over this difficulty, a specific time should be added to ensure the freshness of data. The discovery of data, whether updated or older, becomes effortless in the company of specific counter.

Data Authentication: Data Authentication certifies that information is coming from an authentic source; moreover, it should be verified that no unlawful third party stands in like an authorized user. To attain data authentication, the receiver must have a scheme to guarantee that the received data is approaching from an authentic user. A message authentication code or a hash function scheme can be used to complete data authentication.

Data Availability: WSNs should remain useful to legitimated nodes; data availability ensures that only approved users can access services. One frequent procedure to deal with this problem is additional communication among nodes. A denial of service attack is result of this breach.

Time Synchronization: Time synchronization is critical in wireless sensor network applications; all security schemes should be time synchronized.

Secure Localization: Sensor nodes in WSNs should be localized in a secured manner in any environment. If sensor nodes are not secured properly, advisers can deliver forged position information plus reply messages. There are dissimilar procedures to uncover the genuine location. One of them is ID verifiable, multi-alteration where the accurate location is identified from dissimilar points.

Non-Repudiation: Sensor node insincerity contradicts a message which is sent previously.

Backward Secrecy: In a wireless sensor network, a new joining sensor node should not able to read formerly transmitted messages.

Forward Secrecy: A node which abandons the network is not able to read any impending, potential messages.

1.8 Attacks on Wireless Sensor Networks

WSNs utilize a bulky number of sensor nodes in harsh and inconsiderate surroundings, where sensor nodes are susceptible to malicious attacks. Hence, the data engendered and processed will be supervised by enemies. The objective of the attacker is to dishonestly acquire data and keys stored in neglected nodes. A *strong attacker* is present formerly, everywhere and at any movement, whereas a *realistic attacker* supervises a permanent entitlement of communication after employment. A taxonomy of attacks are separated into two types: passive attack and active attack. In active attack, an attacker endeavors to remove or redesign the messages transmitted on the network. Moreover, the attacker can also introduce their individual messages or react of old messages to disorganize the operation of network. With a passive attack, an attacker is restricted to listening and investigates the interchanged traffic. These types of attackers are difficult to identify, since the attacker doesn't make any changes to the exchanged information. The intention of the attacker is to obtain private information or control of significant nodes in the network [16–22].

Hello Flood Attack: Generally, hello packets are used by numerous protocols to construct a useful hypothesis. Those acquiring such a packet may infer that the dispatcher is surrounded by the radio span of recipient. Attackers make use of a high-power transmitter which persistently dispatches hello packets to fool a huge amount of nodes. On receiving those hello packets, the nodes imagine that the received packets came from within its bordering nodes.

Spoofing Attack: In spoofing, an attack message is altered throughout transmission. Moreover, the attacker alters routing procedures so the data has to pass through a lengthy distance which swallows additional power from sensor nodes. Sometimes the receiver receives erroneous messages. To overcome this problem, we can apply special integrity checking methods like MAC, Hash, etc.

Reply Attack: The attacker interrupts the message and retransmits the same message in the future. Generally, these attackers send a few old messages over the network so the bandwidth of the sensor network decreases. For identifying these old messages, a timestamp or nonce values are used with an encryption algorithm. Among these, the time stamp is preferred because it requires fewer messages.

Sybil Attack: In this attack, one competitor node performs like multiple nodes in order to gain control of the network. The purpose of these malicious nodes is to send bogus messages to overcome the intention of the redundancy mechanism in disseminated data storage systems in peer-to-peer networks. Wireless sensor networks are vulnerable to this attack. For example, a WSN voting system Sybil attack may possibly employ collective competitors to generate supplementary votes. One solution against a Sybil attack is a key registration system.

Wormhole Attack: The most rigorous attack to defend and detect in WSNs is a wormhole attack. A wormhole attack is short concentration link involving two sections of set-up over which an invader replays network messages. That short link could be set off by a unique node or pair off of nodes in special parts of the network that communicate with each other. Adversary nodes capture messages or packets and transmit them to other locations. Wormhole attacks can launched without any information about the network.

Selective Forwarding Attack: In WSNs, data are transmitted from the source to their destination by a multi-hop network connection. Network communication will be good when all nodes need to forward messages exactly. In a selective forwarding attack, one malicious node forwards a message to the nodes' boundary. An attacker may compromise a node in such a way that it selectively forwards some messages and drops others.

Acknowledgement Spoofing: The primary goal of this attack is to make the sender convince the user that a dead or lifeless node is still alive. By doing this, the attacker obstructs communication between sender and receiver. It confuses the sender and misleads the routing process.

Compromised Node Attack: An attacker wants to misappropriate secret information (like security keys) for this attack on some chosen nodes. One resolution is to use a trust protocol, which can make a distinction between compromised and uncompromised nodes.

Node Replication Attack: The attacker obtains secret information like cryptographic keys and copies that key information to another node (called a cloned node). These cloned nodes perform like original nodes and are deployed in network to grab the information of that set-up. The competitor can also introduce forged instructions or influence the instructions passed on through the cloned nodes.

Jamming Attack: In a jamming attack, an attacker interferes by means of the radio frequencies utilized in network nodes. It is absolutely unlike standard radio propagation in the way that it is unnecessary and troublesome, which often results in a denial of service attack. Jamming discontinues the network, which is significant if a solitary node frequency is utilized all through the network. Furthermore, jamming is able to use extreme power at every node by introducing disrespectful packets. Recipient nodes will also swallow energy by receiving those packets.

Tampering Attack: In tampering, an attacker can take out insightful information such as cryptographic keys or other information on the node. The node may also be altered or replaced to create a compromised node. One protection against this attack engages tamper-proof packaging.

Sinkhole Attack: In a sinkhole attack, an attacker renders a negotiated node more interesting and powerful than its bordering nodes to falsify the routing information. As a result, border nodes choose that negotiated node as their following hop to streamline their data thorough the network. Through a sinkhole attack, an attacker can engage almost all the traffic from a particular region.

Traffic Analysis Attack: The purpose of this attack follows as an attacker attempts to examine the traffic to understand the message length, communication delay, message pattern, frequency of communication, and message encoding techniques. Traffic analysis helps to implement other attacks which involve a violation of integrity and confidentiality of messages.

Denial of Service Attack: The aim of the attacker is to create a network arrangement not available to legitimated entities. One common method to implement this attack is to send a large number of false requests to a source network so that network is not able to communicate with the authenticated users in the network. Sensor nodes are mostly affected by this attack because of some senor nodes consists of low-power resources which makes attackers easy to attack with denial of service attack—sending immense figure of fault messages to prepare network unavailable and to not reply intimated services.

Man in the Middle Attack: In this attack an attacker insert or enter itself into the discussion among two networks or parties. An invader starts their personal network connection by entities and conveys memorandum between them. The invader then gains the entities' acceptance, intercepts all incoming messages, and sends new messages.

1.9 WSN Standards

The IEEE defines various standards for each individual network in communication technology [23]:

- The IEEE defines 802.11 which indicate Wi-Fi Service. It characterizes the diversity of standards, which concentrate on the security, the superiority of services, and other issues.
- IEEE 802.11a defines a wireless sensor network possessor working on 5 GHz ISM band with data communication extent until 54 Mbps.
- IEEE 822.11b defines a wireless sensor network occupation on 2.4 GHz ISM band with a data rate as far as 11 Mbps.
- IEEE 802.11e covers prioritization and QoS.
- IEEE 802.11f handles networking handover.
- The IEEE 802.11g standard is wireless sensor network commission occupation on 2.4 GHz ISM bands; data communications fit to 54 Mbps.
- IEEE 802.11h covers energy management in transmission.
- IEEE 802.11i defines authentication and encryption technologies for secure transmission.
- The IEEE 802.11j standard able to connect, communicate or exchange data; 802.11k defines system measurement report.
- IEEE 802.11n operates at 2.4 and 5 GHz ISM bands with data communication or transmission capable of 600 Mbps.
- IEEE 802.11s, 802.11ac, 802.11ad, 802.11af are used for mesh networking, multi-station operation, high throughput, and white Wi-Fi.
- The first accepted wireless standard was 802.11b and 802.11n standard was the newest one which operates at 2.4 GHz and its data rates up to 60 Mbps.

1.10 Routing Protocols in WSNs

In wireless sensor networks, one of the most important sensor nodes sensed data must be transmitted to the base station in a multi-hop network with a minimized energy consumption level. There are many routing protocols proposed: some are based on the location information of the sensor node (called *geographical routing*) and some are based on *location-aware routing* protocols [10,14,15,24–29].

1.10.1 Sensor Protocols for Information Negotiation

SPIN (sensor protocols for information negotiation) hosts a family of protocols which come with special types of impression and characteristics. SPIN is called a flat-based routing algorithm, which means that it assumes that all sensor nodes play an equal role in the network. SPIN family protocols are used to deal with a shortage of flooding and gossiping. SPIN is an adaptation of flooding; in classic flooding, every message or packet transmits to every affectionate link of the node which takes draining out the battery life. So, spin protocol creates metadata which contains description of original data. When a node senses the data, it broadcast metadata to its surrounding nodes. If some surrounding node is given an appeal or request, it is sent to the source and the source sends the original data. SPIN utilizes three types of messages: (1) ADV (advertisement) for promotion fresh data, (2) REQ (request) for sending a request for data, (3) and DATA (generative data) for data messages containing original data.

When a node has data to distribute, it announces it by broadcasting an ADV message. The ADV message contains only metadata which describes the type of data enclosed by a sender node, like the volume, contents, and essentials of the message. Nodes which receive the ADV message check whether there is any useful information in that message. If it's interesting, then the node sends a REQ message to the node that advertises a metamessage. Here, REQ contains metadata again. At last, the advertised node sends the DATA message which contains original information with a metaheader to the requesting sensor node. Figure 1.4 shows a node 3 broadcast ADV to all adjacent nodes 1.2 and 4. After verification of the metadata, node 4 requests for the original DATA using REQ. Then node 3 sends the message DATA to node 4. After getting DATA, node 4 also sends an ADV to the nearest nodes, 5 and 6, and this process continues. Node 4 does not send ADV to node 3, since it recognizes that the data is received from node 3 [14,26].

1.10.2 Low Energy Adaptive Clustering Hierarchy

The low energy adaptive clustering hierarchy (LEACH) is a hierarchical protocol, and nodes in the network are not useful when the batteries goes

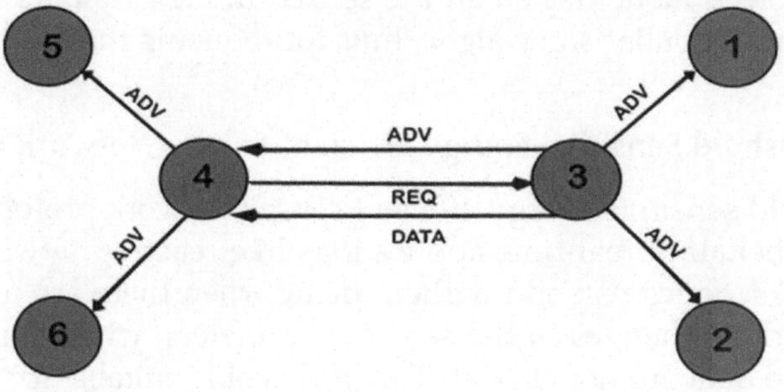

FIGURE 1.4
Working principle of SPIN protocol.

down. In WSN data transmission, the starting sensor node to the base station is straightforwardly considered as a costly operation. LEACH increases the duration time of nodes by allocating the simplest method of transmitting the data. So, a few nodes elected as clusters, and those collect and forward the sensed data to the base station. LEACH algorithms function on distinct phases (see Figure 1.5); the first phase is an advertisement phase in which each sensor node has the potential to become a cluster head, depending on the sensor node with maximum energy and if it wants to become as a cluster or not. The second phase *is set-up phase* in which a sensor node makes a decision about its cluster head, depending on the signal strength received in the advertisement. The third and final phase is the *data transmission state*, in

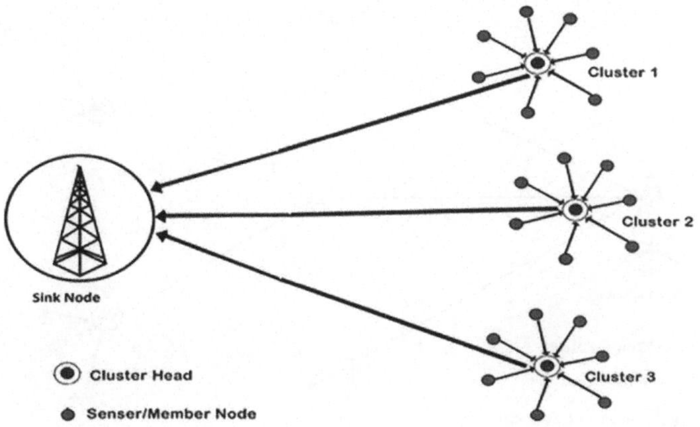

FIGURE 1.5
Working principle of LEACH protocol.

which each sensor node sends sensed data to the cluster head, except when they are in sleep mode. When all the sensed information is received, then the cluster head applies some algorithms for reducing the data [15,27,28].

1.10.3 Threshold Sensitive Energy Efficient Sensor Network Protocol

The threshold sensitive energy efficient sensor network protocol (TEEN) is suitable on behalf of real-time applications like reactive networks. In reactive networks, nodes respond without delay when there are unanticipated and tremendous changes in the sensed parameters, which also save node energy if the data are not critical. This protocol is suitable for time-critical applications. TEEN uses two threshold values to measure the data, namely hard and soft threshold values. The hard threshold value is fixed ahead of the sensed characteristic. If the node senses a hard threshold value it exchanges on its transmitter as well as sends it to its cluster head. The soft threshold value is a minute variation in the value of sensed characteristic, which prompts the node to exchange its transmitter and pass it on. Sensor nodes discern their location endlessly if the primary time a parameter from the attribute set accomplishes its hard threshold value. The node exchanges the transmitter and sends the sensed data. This is termed as *sensed value* and it is stored as an internal variable in the node. The node throws more transmitted data at this cluster head time when the modern value of sensed attribute is superior than hard threshold value. The present sensed attribute differs from the stored value by an amount alike or superior than the soft threshold [29]. Figure 1.6 shows the basic working principle of the TEEN protocol.

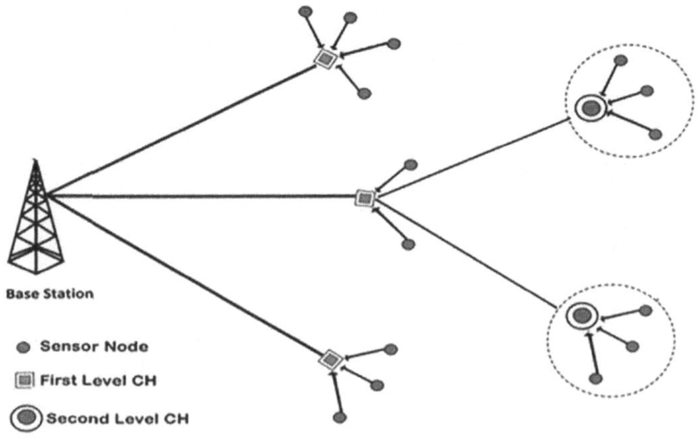

FIGURE 1.6
Working principle of TEEN protocol.

1.11 WSN Simulations

The exponential growth of wireless sensor network research has raised testing issues in the micro-behavior of proposed protocols and their performance estimation. The techniques for performance evaluation of proposed protocol algorithms in wireless sensor networks are simulation, mathematical modeling, and real experimentation. For experimentation, the process cost and time are high and mathematical models get complicated which provide a limited insight into the system design. Even though simulation is the best choice for WSN research, it has some common faults, like inappropriate details, awareness of simulation, features, etc. [30,31].

Here, we provide minute explanations about the simulation tools of WSNs:

Tiny OS: A tiny OS simulator (or TOSSIM) is a discrete event simulator specially designed for WSNs. TOSSIM is free and consists of a FIFO task scheduler. Software components, such as memory, radio communication, sensing, and other devices, are also included. The ingredients of TOSSIM structural design consist of a network topology graph, support for compiling, a discrete event queue, and simulated hardware components (like radio stack, sensors, and communication infrastructure) that are in touch with exterior programs. TOSSIM contributes a run-time configurable debugging output in which the implementation of a request can be monitored from special viewpoints without recompiling it. For visualization plus control of the simulation, it runs on TinyViz, which is a Java-based graphical user interface. The TOSSIM simulation atmosphere make available mechanisms to cooperate with network, monitor packet traffic, and inject packets into the network. The TOSSIM framework effectively makes use of the testing and analysis purpose.

COOJA: The contiki operating system for Java is also known as COOJA. It's a specially designed sensor network simulator for the Contiki OS. C and Java are programmed for the COOJA simulator, which offers a quick and easy way to develop new application modules.

NS2: Network simulator-2 (NS2) is a discrete, event-based, open source, and REAL (known as progressive network) network simulator specially designed for sensor nodes, created by University of California and Cornell University. NS2 contains two languages C++ and OTcl (object-oriented tool command language). C++ defines the simulation objects and OTcl accumulating, and constructs the objects as forecast distinct proceedings for supplementary simulation. Attackers build up their own C++ objects with use of an OTcl configuration interfaces to put together these objects. To check a large number of dependencies, this further leads to outsized computation time plus memory requirements. Traces and Monitors are utilized to accumulate data where log events are connected to a packet broadcast required for computing performance constraints, like throughput or traffic delay.

NS3: To provide better debugging and configuration, NS2 is upgraded as network simulator-3 (NS3). NS3 simulation modules implementation provides close to actual level implementation. When moving from simulation to experiment, it reduces discontinuous values. Its structural design uses C++-based representations for easier debugging; furthermore, a Python-supported scripting API is amalgamated with additional Python-supported environments or programming models. NS3 simulation representation programs can be written as either C++ main () programs or Python scripts; it is a free and open source project.

OMNET++: Objective Modular Network Test-bed in C++ (OMNET++) is a component-based simulator. OMNET components are programmed in C++. OMNET++ models are a collection of hierarchal, nested modules. System modules consist at the top level, which contains sub-modules each of which further contains other sub-modules. It eradicates interdependency among NS2 simulator modules which, in turn, make use of memory proficiently, diminishing software difficulty with a quicker response time compared to NS2. Simple modules are implemented by users in C++ via an OMNET++ simulation class library. Messages with distinctive IDs are employed for communication among modules. Messages represent frames or packets in a computer network; the local reproduction time proceeds when a module obtains messages from any more modules or from itself.

Avrora: Avrora designed to implement cycle-accurate simulations, event queues, multithreading, and thread harmonization; at each thread it runs one node. Avrora is an instruction-level simulator written in Java that helps to achieve flexibility and portability. The thread organization guarantee is that global timing, along with radio communication, are protected throughout the simulation.

ATMEU: ATMEU (Atmel EMUlator) is a software emulator for AVR processor-based systems, including support for peripheral devices like radio on the MICA2 sensor node. It provides opportunities for setting breakpoints, watch points, and single stepping for multiple nodes in a sensor network. ATMEU makes use of the binary record that is loaded onto the target MICA2 node. Furthermore, the AVR processor emulation engine excessively prompts model code implementation on every sensor node, so an undersized number of information of the authentic operation of sensor nodes are abstracted out, and thus endow it with reasonable results.

J-sim: Java-sim (J-sim) is an open source, component-based, generic simulation framework. Implementation in Java, with autonomous components makes it independent, reusable, and extensible. J-sim can be integrated with different script languages like Perl, Tcl, and Python. J-sim is put into action on top of autonomous component architecture (ACA) which is component-based software structural design. ACA components are exchanged with a few words by distribution and getting data on their ports.

1.12 WSN Operating Systems

Wireless sensor networks function mutually at the node level as well as the network level. Sensor nodes at the network level convene with communication and connectivity. However, at the node level, sensor nodes are restrained with limited power sources, processing, memory, and radio frequency, which are vital for sensor networks to claim for connectivity and communication. Other than this, the operating system (OS) needs to protect operation modes of CPU, providing an interface to user applications. The purpose of the operating system is to bear the expansion of dependable software applications through a well-situated and nonviolent extraction of hardware. For sensor nodes, the OS obscures squat-level particulars of detecting, given that interface to the application layer. There are some challenges to designing operating systems for WSNs, such as a miniature memory footprint; proficiency in power plus computation; lack of hardware parallelism, real-time, and robust delivery in application and design. Some OSs for sensor nodes are listed below [32,33]:

Monolithic: Monolithic systems provide a sophisticated simulated interface above hardware by established primitives to operate the entire OS. Monolithic systems don't follow any detailed structure, they provide a straightforward kernel. The entire operating system runs in kernel space in monolithic OSs.

Microkernel: Microkernel operating systems provide a more structured architecture than conventional and monolithic. Microkernels provide important amenities implemented through a set of procedures, which in return grant the advanced level OS functionality (such as thread management, address space, and interprocess communication). All other services are provided by kernels, such as interaction applied in userspace agendas denoted as servers. The real goal is to make microkernel OSs as small and simple as possible. Moreover, modularity is key to the design operating system.

Event-Based OS: An event is a self-determining accomplishment which takes place inside the system or interacts with other system events. The event is signaling obsessed and can be prompted by a purpose which contests an event. It is initiated individually when it's receiving a prompt indication from an interrupted service routine. This kind of OS is suitable for WSNs because fewer properties and resources are required in additional energy competent systems. Also, event-driven OSs has advantages, such as a low-price scheduling procedure and extreme transportability; as well as some disadvantages, such as an event loop in protocol, a schedule that needs to be cut to subgroups, and restricted buffer producer–consumer problems.

Thread-Based OS: Thread OSs contains a minimum single thread for single process termed as main thread; as soon as the mother process thread is formed at the time the single process threads are created. The process makes an increased OS effective throughout the amended consumption of core memory and decreases the computation of properties. Kernel operation is achieved via a set of thread processes. Some threads have real-time requirements; they need to respond quickly and correctly. Threads are similar functions, a respective thread by its individual stack includes thread control block. Advantages of thread-driven OSs are abolish-restricted buffer problem, a programmer in charge of the program, self-regulating preparation, simulated parallel implementation. Some disadvantages are complex shared memory, widespread context switches, complex stack examination, and soaring transportables owed to stack management.

1.13 Challenges of WSNs

Some designing challenges for WSNs are listed below [2–6,12,34,35]:

Fault Tolerance: Depending upon the application, the sensor node may increase to hundreds to thousands of nodes. Due to physical damage or environmental changes, a few sensor nodes may possibly stop working or become blocked. Failure nodes must not change the whole mission of the sensor network; this is the trustworthiness of the fault tolerance problem. The capability to prolong sensor network functionality, not including whichever intermission payable to this sensor node failure, is called *fault tolerance*. There are plenty of fault tolerance and energy proficient algorithms that can be executed to enlarge the lifespan of the network. So, destroying one node doesn't affect the network performance. For example, consider some sensor nodes deployed in-house to track temperature and humidity; the toleration value must be small. Therefore, only this breed of sensor network cannot simply be scratched through territory noise. According to numerous appliances, the death of the foremost node is considered as the existence time period of the sensor network, despite the fact that in some occurrences, a network can stay alive until the death of the final node.

Production Cost: This broad constraint in sensor network design is to provide the lowest sensor node cost as possible. WSNs consist of a great number sensors to calculate or estimate network cost, while the outlay of the particular sensor node is more important. With the outlay of one sensor node, we can buy 10 Bluetooth devices, which are known to be low-cost devices. Take a look at the sensor node in Figure 1.2, which consists extra units like sensing unit, a processing unit, a location finding system, and a mobilizer which makes the sensor node more expensive. So, the cost of the sensor node becomes as challenging with all the application requirements.

Scalability: Sensor nodes might at any time monitor an operation in the thousands. The number may rise to tens of thousands to complete the application. Adding some other nodes to the already existing network doesn't affect the continuous performance of sensor network. The quantity of nodes in a particular area are used to indicate the nodes' density value, where densities depend on node function and the sensor nodes are deployed.

Network Topology: The geographical design of a sensor network is called network topology; it has to enclose a high-density value. A wireless sensor network mainly consists of three topologies, namely pre-deployment, post-deployment, and re-deployment. *Pre-deployment* is an initial phase of sensor node deployment which is done by throwing or placing one after another in the network field. This type of deployment places nodes one by one either manually or mechanically (released from plane, rocket, missile, etc.). Insertion in the factory, even though these sensor nodes are deployed according to the engineering measurement, are the schemes used for initial deployment nodes in the network. There is handsome compensation for using the above methods; promoting self-organizing and avoiding fault tolerance, abolishing the need of any pre-deployment, and re-planning generally diminishes the initialization cost. *Post-deployment* occurs after initial deployment of sensor node topology, which may change due to some significant operation on sensor network; jamming, sensor nodes moving from one place to another to complete the assigned task so the topology changes may also change vacant energy, reachability, task details, and malfunctioning. Device failure is common in any network design due to energy shortage or some other demolition. Thus, sensor networks' geometrical and spatial relations are horizontal towards repeated challenges behind deployment. *Re-deployment* is used when some sensor nodes do not work properly which may be altered by changes in the environment. So, some auxiliary sensor nodes can re-deployed at the place of shattered nodes or appropriate changes in assignment dynamics at any time. The addition of fresh nodes may necessitate reorganizing the network.

Data Delivery Mode: Wireless sensor network data is mostly categorized into; reactive, proactive, hybrid, query-driven, and event-driven; among these elect, one data delivery model is needed in the application of wireless sensor networks.

Global Identification: Wireless sensor networks have a bulky amount of sensor nodes in the set-up which does not hinder global identification. A GPS grants location information to nearby sensor nodes, however it principally requires line-of-sight from several satellites. But, in some cases, it is not possible because of sensors placed inside a building, underwater, or below dense undergrowth, or those that become jammed by eavesdroppers.

Storage and Retrieval: Sensor networks study the environmental phenomena and fabricate a large amount of raw data in a continuous time series. While data is continuous, traditional database management is not suitable for the wireless sensor networks. Three types of memory are utilized in wireless sensor nodes: flash memory, fuse bit, and electrically erasable programmable read-only memory (EEPROM).

Area Coverage: WSNs are constructed with limited number of sensor nodes. Area coverage tends to spread across the entire region with a minimum quantity of homogeneous sensor nodes. Classically, WSNs consists of two key processes: (1) area coverage and (2) computational complexity.

The first one is after organizing discretely in pre-defined section. Sensor nodes account for environmental facts endlessly. Furthermore, sensed substances are sent to a base station by the use of a communication component, either straightforwardly or through additional nodes, even though this category of sensor uses a lot of energy. We know that the broadcast charge is higher than the computation charge in WSNs. First-rate area coverage optimization procedure in WSNs accomplishes enhanced exposure superiority within a smaller computational time. Improved cuckoo search is one of the evolutionary optimization algorithms; modeled after the cuckoo bird which lays its eggs on the nests of other bird species. If the entertainer or host bird is proficient enough to figure out the intrusion, it will either pitch the alien eggs aside or pitch out its solitary nest and assemble a fresh one. Virtual force-based coverage optimization, pure genetic algorithm, Monte Carlo GA, optimal genetic algorithm, improved GA, particle swarm optimization, Democratic PSO, and artificial bee colony are some other algorithms for better optimization with area coverage.

In-Network Processing: In wired and wireless networks, transport protocols are used to transmit data from end to end, guaranteeing that the data submitted from senders to receivers are not modified by intermediate nodes until they reach their destination or receiver. Wireless sensor network data can be modified or aggregated by intermediate nodes in order to remove the redundancy of information.

Latency: Latency refers to the delay of packet transmission from the sender until the packet reaches the receiver's side successfully. Sensor data is valid only temporally because the environment changes rapidly. Therefore, it is important to receive sensed data in a timely manner without any latency.

Network Lifespan: The sensor node is a small electronic device with limited resources and energy; as a result, these networks have a limited lifespan. A network is considered as effective when all of its nodes are in an inactive state. When networking in the effective state, it can monitor the entire area, as well as accumulate the sensed facts through a predefined superiority of services. Suitable procedures attempt to reduce energy usage and increase the lifespan of the network.

References

1. I.F. Akyildiz, W. Su, Y. Sankarasubramaniam. "Wireless Sensor Networks: A Survey, Computer Networks." *Elsevier BV* 38 (2002), 393–422.
2. Sanjeev Kumar Guptha, Poonam Sinha. "Overview of Wireless sensor networks: A Survey." *International Journal of Advanced Research in Computer and Communication Engineering* Vol. 3, Issue 1, January 2014.

3. Neha Singh, Kamakshi Rautela. "Literature Survey on Wireless Sensor Networks." *International Journal of Engineering and Computer Science* ISSN: 2319–7242, Vol. 5, Issue 8, August 2016, 17544–17548.
4. Luís M.L. Oliveira. "Wireless Sensor Networks: A Survey on Environmental Monitoring." *Journal of Communications* Vol. 6, No. 2, April 2011.
5. Bimlendu Shahi, Sujatha Dahal, Abhinav Mishra et al. "A Review Over Genetic Algorithm and Applications of Wireless Sensor Network Systems." *Elsevier BV,* (2016), 431–438.
6. M.S. Sunitha, Jayoti Malik. "Comprehensive Study of Applications of Wireless Sensor Network." *IJARCSSE* Vol. 2, Issue 11, 2012, 56–60.
7. P. Bauer, M. Sichitiu, R. Istepanian, K. Premaratne. "The Mobile Patient: Wireless Distributed Sensor Networks for Patient Monitoring and Care." IEEE proceedings EMBS International Conference on Information Technology Application in Biomedicine, 2000, pp. 17–221.
8. Dey Nilanjan, Amira S. Ashour, Fuqian Shi. "Developing Residential Wireless Sensor Networks for ECH Healthcare Monitoring." *IEEE Transactions on Consumer Electronics* Vol. 63, Issue 4, 2017.
9. Dey, Nilanjan, Amira S. Ashour, Fuqian Shi, R. Simon Sherratt. "Wireless Capsule Gastrointestinal Endoscopy: Direction-of-Arrival Estimation Based Localization Survey." *IEEE Reviews in Biomedical Engineering* 10 (2017): 2–11.
10. I.F. Akyildiz, W. Su. "A power aware enhanced routing protocol for sensor networks." Georgia Technical Report, January 2002.
11. Shabbir Hassan, Md. Zair Hassan. "A Survey of Wireless Sensor Networks." *International Journal of Emerging Technology and Advanced Engineering* Vol. 3, Issue 3, March 2013.
12. Ridha Soua, Pascale Minet. "A Survey on Energy Efficient Techniques in WSN." IFIP, WMNC 2011.
13. Louis M. Borges, Fernando J. Velez. "Taxonomy for Wireless Sensor Network Services Characterization and Classifications." University of Beira, IST-UNITE, 2007.
14. Adrian Perrig, Robert Szewczyk et al. "SPINS: Security Protocol for Sensor Networks." Mobile Computing and Networking, Italy, 2001.
15. ReshamaI, Tandel. "Leach Protocol in Wireless Sensor Networks: A Survey." *International Journal on Computer Science and Information Technology* Vol. 7, Issue 4, 2016, 1894–1896.
16. Mohammad, Lamine Messai. "Classification of Attacks in Wireless Sensor Networks." *International Congress on Telecommunications and Applications,* Abderrahmane Mira University–Bejaia, Algeria, 23–24, April 2014.
17. Jaydip Sen. "Security in Wireless Sensor Networks." Department of Computer Science & Engineering, National Institute of Science & Technology, India.
18. Zinaida Benenson, Peter M. Cholewinski, Felix C. Freiling. "Vulnerabilities and Attacks in Wireless Sensor Networks." Laboratory for Dependable Distributed systems, University of Mannheim, Mannheim, Germany.
19. Madhumitha Panda. "Security in Wireless Sensor Network Using Cryptographic Techniques." *American Journal of Engineering Research* Vol. 3, Issue 1, 50–56, 2014.
20. Rishav Dubey, Vikram Jain. "Attacks in Wireless Sensor Networks." *International Journal of Scientific and Engineering Research* Vol. 3, Issue 3, March 2002.

21. "Security in Wireless Ad-Hoc Networks: A Survey." Computer Communication Elsevier, 2014, Vol. 51, 1–20.
22. A. Perrig, J. Stankovic, D. Wagner. "Security in wireless sensor networks." ACM Vol. 47, No. 56, June 2005.
23. Nandini Mukharjee, Sarmistha, Sarbani Roy. "Wireless Sensor Network Architecture." Building Wireless Sensor Networks, Taylor & Francis Group, CRC Press, 2016.
24. Luis Javier García Villalba, Ana Lucila Sandoval Orozco, Alicia Triviño Cabrera et al. "Routing Protocols in Wireless Sensor Networks." Sensors, ISSN 1424–8220, 2009.
25. K. Akkaya, M. Younis. "A Survey on Routing Protocols for Wireless Sensor Networks." *Ad Hoc Networks* Vol. 3, No. 3, 2005, 325–349.
26. Debao Xiao, Meijuan Wei, Ying Hou. "Secure SPIN: Secure Sensor Protocol for Information via Negotiation for Wireless Sensor Networks." IEEE Conference on Industrial Electronics and Applications, May 24–26, 2006.
27. Li Yuling, Feng Liu. "The improvement of LEACH Protocol in WSN." Computer Science and Networking Technology ICCSNT 24–26, December, 2011.
28. Reenkamal Kaur Gill, Priya Chawla et al. "Study of LEACH Routing Protocol for Wireless Sensor Networks." International Conference on Communication Computing and Systems, 2014.
29. A. Mmajeshwar, D.P. Agarwal. "TEEN: A Routing protocol for enhanced efficiency in wireless sensor networks." Parallel and Distributed Processing Symposium, *IEEE Proceedings*, August 7, 2002.
30. Ankita Sharama. "A Survey on Wireless Sensor Network Based Approaches." *International Journal of Advanced Research in Computer Science and Software Engineering* Vol. 4, Issue 4, April 2014.
31. Sachin Garjjar, Mohanchur Sarkar, Kankar Dasgupta. "Wireless Sensor Network Simulation." *Journal of Electronic Design Technology* Vol. 5, Issue 2, 2014.
32. Muhammad Omer Farooq and Thomas Kunz. "Operating Systems for Wireless Sensor Networks: A Survey." Sensors 2011, 11, 5900–5930; doi:10.3390/s110605900.
33. Girts Strazdins, Atis Elsts, Krisjanis Nesenbergs, Leo Selavo. "Wireless Sensor Network Operating System Design Rules Based on Real-world Deployment Survey." *Journal of Sensor and Actuator Networks*, ISSN 22224-2708, 2013, 2509–2556.
34. Huynh Thi Than Binh, Nilanjan Dey, Nguyen Thi Hanh. "Improved Cuckoo Search and Chaotic Flower Pollination Optimization Algorithm for Maximizing Area Coverage in Wireless Sensor Networks." *Neural Computer Applications*, Springer, 2016.
35. Chee Yee Chong, S.P. Kumar. "Sensor Networks: Evolution, Opportunities and Challenges." IEEE Proceedings, Vol. 91, Issue 8, 2013, 56–60.

2

Optimization Problems in Wireless Sensors Networks

D. K. Sah, Chaya Shivalingagowda, and D. Praveen Kumar

CONTENTS

2.1 Introduction

Wireless sensor networks have gained worldwide attention, especially with the micro-electro-mechanical system (MEMS) technology, which helps in the development of smart sensors [1]. It incorporates a large number of multifunctional sensors to form the network. The prime concern of WSNs is energy because of the limitation of the power supply. In this context, the geostationary location of the applications vary and are not accessible in most scenarios, creating further hurdles in the replacement of the power source (typical sensors are battery powered). These limitations can be compensated through careful optimization of the resources. There are many stringent resources (apart from energy) in WSNs such as bandwidth, memory and QoS (end-to-end-delay, packet latency, reliability) which can be solved by an optimization algorithm. Optimization methodologies [2–5] have typically been adapted in WSNs to enhance the network lifetime. The application area of WSNs is very broad and the classification is shown in Figure 2.1, based on

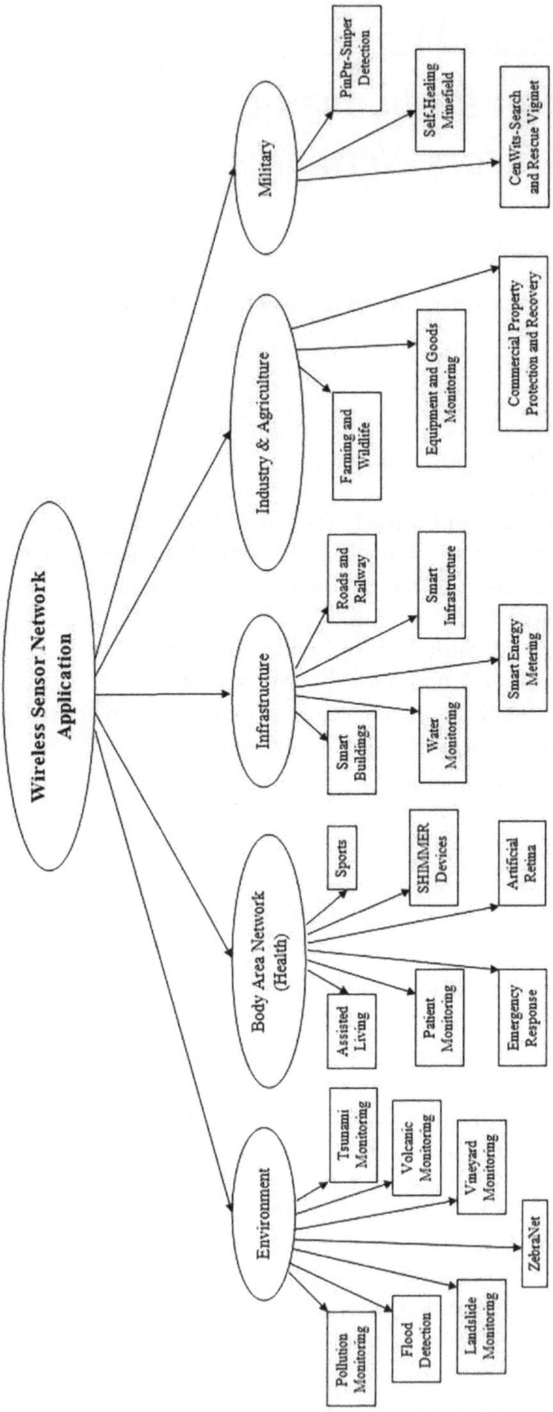

FIGURE 2.1
Classification of WSN applications.

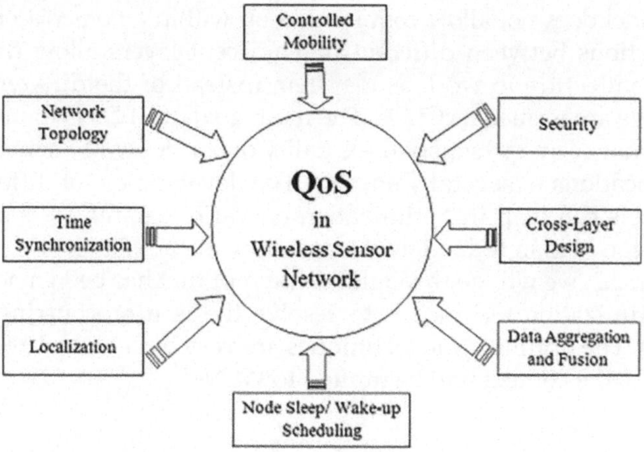

FIGURE 2.2
Performance matrices dependability on QoS in WSNs.

observation. Energy plays a vital role in WSNs, and can surely compensate through the QoS, therefore in Figure 2.2 the dependability of performance matrices on QoS in WSNs is provided.

In this context, the optimization of resources (objective function) means to get the input followed by the constraint, which will supposedly bind the solution with the availability of the resources. To perform the optimization of resources, we require objective functions, optimization variables, constraints (optional) and methods to quantify and compute objective functions bound by the constraints (optional). Some algorithms, which need an account to provide the solution, require subsidiary variables that will be covered later in the text. Typical communication protocols follow the modular architecture in which each module or layer is preconfigured to perform a particular task; such as MACs responsible for sensor node scheduling, network layers for data forwarding from sensors to sink and a physical layer for a modulation technique. The OSI model is common in communication networks and provides a networking framework to implement the protocol in seven layers. Apart from that, some other protocols (such as TCP/IP, ZigBee and so on) provide an abstract modularity of layers which are followed as a typical WSN protocol architecture. Apart from the modular information sharing between the layers, another architecture aspect called cross-layer designs (CLDs) is also popular for performing the task. It helps performance by increasing the availability of the information (operational statistics need protocols) among the layers. In Chapter 2, the taxonomy of the optimization in WSNs will be examined in the context of the architectural design of the protocol stack. Cross-layer communication, architecture can be defined as 'any attempt, which hinders the encapsulation in the functionality of the modules/layers', will be categorized as a CLDs. Encapsulation of layers in the OSI model does not allow communication within a non-adjacent node [6]. The interactions between different non-adjacent layers allow the communication architecture to work as a system instead of the different protocol stacks. These are termed as CLDs. The main goal of CLDs is to improve system performance and to satisfy the Quality of Service (QoS) demands of different applications in several scenarios. The development of different CLDs and protocols can help in optimization on various parameters at different layers and can help in reduction of transfer overhead [7].

In Section 2.2, we will go through the attempt that has been made through several optimization techniques to resolve the issues occurring in WSNs (Figure 2.2). The optimization techniques are very broad in nature, therefore in Chapter 2, the context will be limited to WSNs.

2.2 What is Optimization?

In science and technology, optimization deals with the efficient use of available resources with desired and undesired factors, accounted for in such a way so that the best performance can be achieved by the system. In the context of WSNs, resources (such as the energy of nodes, frequency and the memory of sensor nodes) and processing capabilities of the individual network are very precious and must be used efficiently. Typical optimization problems consist of the objective function followed by the constraints associated with it. The objectivity of the function is to minimize or maximize the objectives, which can be classified as single-objective functions and multi-objective functions. The constraints associated with the forms of the variables are used to define the boundary of the solution to a range of the solution. The formal definition of the optimization problem (minimization) having n variables and $m \geq 1$ objectives can be defined as:

$$\min f(x) \quad // \textit{single objective} \tag{2.1}$$

$$\min\left[f_1(x),\ f_2(x),\ f_3(x), \ldots\ f_m(x)\right] \quad // \textit{Multi-objective} \tag{2.2}$$

$$\text{Subject to} \qquad h_i \leq 0 \ i = 1,2\ldots.m_i \tag{2.3}$$

$$k_j \leq 0 \ j = 1,2\ldots.m_j \tag{2.4}$$

where x and $f(x)$ can define a decision space and objective space, respectively. The variables h_i and k_j are constraint-associated. In the context of optimization problems, the properties and derivatives of the variable and equations used to make further categorization of optimization problems, such as linear, non-linear, quadratic, integer-lineal, etc. for single and multi-objective problems. A typical optimization process is the combination of the following three components [8]:

1. The Models

   Further classification of a model a can be possible which lead either to the mathematical model or numerical model; this solely depends upon the application scenario.
2. The Optimizer/Algorithm

   This has several options available based upon the dependencies and the optimizer and can be further classified under population, trajectory, derivative-based and derivative-free, stochastic, memory-less, history-based and bio-mimic categories, among others.
3. Evaluator or Simulator

   The solution proposed by the optimizer or algorithm also leads to some classification which indicates the methodology for obtaining the solution. The evaluation can be possible in many ways, such as direct calculation, numerical simulation, experimental trial and error, or some combination of the above.

Meanwhile, the scope of Chapter 2 is limited to the optimization methodology in WSNs. Therefore, we encourage the readers to see [8–10], for a better understanding of the optimization model.

2.2.1 WSN Optimization

When optimizing WSNs, it is necessary to consider both the problem and the solution. The type of problem and solution will depend on factors such as the optimization goal or network model.

a. WSN Optimization Problems

   In the context of WSN, optimization can be applied at different design levels. This approach helps meet the custom requirements that depend on the purpose of the application [10]. Most approaches in the literature focus on network and data link-level optimizations, defining MAC and routing protocols, respectively. The protocols of MAC typically target throughput, load-balancing and energy consumption optimization; whereas routing protocols usually address data delivery delays, query dissemination and network topologies. Apart from network protocols, a different type of problem is optimal sensor deployment [8]. The goal of the WSN deployment problem is to ensure full coverage and connectivity with the minimum number of sensors in an area of interest. As explained in [8], this problem is NP-hard (non-polynomial hard). Coverage depends on how well the sensors are monitoring the environment or moving objects in the area where the sensors deployed. In this context, each of the nodes need to know the shortest path between every point on the path and its nearest sensor node, which is also NP-hard.

b. WSN Optimization Solvers

There are many general strategies to deal with optimization problems and some others explicitly designed for WSNs. In [9,11] the authors proposed an optimization approach intended for WSNs. They classified these problems into a new category called F-Lipschitz, which claims to be fast and based on the Lipschitz property of constraints. In particular, the optimization problem requires an increased objective function and the constraints transformed into contractive Lipschitz functions. By casting a WSN optimization problem as an F-Lipschitz problem, the authors showed that this form is robust and not sensitive to perturbation of the constraints and quantization errors, which are quite essential for WSNs with low computational precision. Moreover, F-Lipschitz optimization solves problems traditionally approached by Lagrangian methods more efficiently (a Lagrangian method is a problem transformation that redefines the objective function using the problem constraints, determining how much the optimum is affected by each of the problem constraints). Among others, the authors proposed F-Lipschitz for distributed detection and radio power allocation.

Other familiar nonlinear problem solvers are genetic algorithms (GA) [12]. A GA is a global search algorithm that mimics the process of natural selection using techniques inspired by genetic mechanisms. Based on the biological environment, the optimization problems are modelled. In an optimization problem, each chromosome in a population treated as a possible solution. The fit and strong chromosomes only live after some rounds of evolution and the weak chromosomes are discarded. In [13], the authors proposed an optimization structure for collaborative and distributed beamforming in a WSN. The objective of the optimization is to reduce the side-lobe level. In this case, the complexity the system constraints and of the objective function make the utilization of GA a reasonable option. Bio-inspired intelligent optimization has also proposed in the WSN arena [14].

In [15], the authors proposed an improved cuckoo search and chaotic flower pollination to a solve area coverage problem for heterogeneous wireless sensor networks with a given set of sensor nodes. These two proposed algorithms have reduced the computational complexity while providing efficient coverage when compared with existing approaches.

In [16], the authors have surveyed a localization method based on the direction of arrival for wireless capsule gastrointestinal endoscopy.

Bio-inspired optimization can be understood as 'survival of the fittest' applied to a WSN. One of the most well-known frameworks from these is ant colony optimization, which is motivated by the searching behaviour of ants to collect food. There are many applications solved by using this approach, such as graph colorings, the traveling salesperson problem, routing in the network, etc. Ant colony optimiazation is an NP-hard problem, the worst-case time complexity is $O(2^n)$, because of the distributed and decentralized nature of ants in foraging between the source and destination. Another well-known, population-based search problem was particle swarm optimization (PSO) [17,18], where each particle was defined by a potential solution of a problem in a D-dimensional space. With the ith particle represented as $X_i = (X_{i1}, X_{i2},\ldots,X_{iD})$, each particle adjusts its position close to the optimal according to its own experience and that of neighboring particles. There is a large family of useful, PSO-based algorithms to solve non-convex, nonlinear optimization problems (a function considered convex in an interval if it is differentiable within the range and has only a local minimum).

2.2.2 Significance of Optimization in WSNs

In Table 2.1, categorization was based on different performance metrics of the optimization approaches in the given WSNs. To generalize these in the context of WSNs, the objective of minimizing energy consumption or the maximization of a network lifetime is dependent on various parameters such as latency, packet error, coverage, connectivity and so on. Such interdependability of the performance matrices makes the resource allocation process difficult to optimize the objective. Therefore, several authors have taken different optimization models (based on their intentions) to achieve their goals in WSNs.

TABLE 2.1

Classification of Existing Approaches of Optimizing Methodology

Objective	Reference	Optimization Approach	Result Analysis
Coverage	[19–23]	Mathematical programming, hybrid algorithms	Simulation
Network lifetime	[19,24–29]	Protocol design, mathematical programming, hybrid algorithms, EAs and theoretical analysis	Simulation
Energy consumption	[21,30–33]	Protocol design, mathematical programming, hybrid algorithms, EAs, SIOA and theoretical analysis	Simulation
Efficiency	[34–37]	Protocol design, hybrid algorithms and SIOA	Simulation
Network latency	[31–33,38,39]	Protocol design, mathematical programming, hybrid algorithms, EAs, SIOA and theoretical analysis	Simulation

2.2.3 Protocol Architectural and Its Significance in WSNs

In the OSI model, each layer has its definitive predefined functionality for communication and allows for interaction or procedure calls between the adjacent layers, but does not allow interaction in between the non-adjacent layer. The working principle of a traditional layered protocol based on an OSI model or TCP/IP model has initially been designed for wired communication, and provides an abstract view of each layer as it will be able to provide services and information exchange between adjacent layers [13,40–43]. The nature of encapsulation provided by traditional layers offers modularity in network deployment. Therefore, it also prevents sharing some necessary information among the layers in the protocol stack, which creates a hindrance in many applications and causes a negative impact on the efficiency of the network due to its nature of abstraction. For example, a medium access control (MAC) senses the neighbor node, which is supposed to forward the data, but is highly occupied due to traffic load. But again and again sensing for that particular node to free instead for data forwarding [44], it became highly desirable to propose a cross-layer (CL) protocol stack to overcome the limitation and provide better network management in terms of energy, QoS, and other aspects such as security and multimedia data handling with optimal latency.

The interactions between different nonadjacent layers allow the communication architecture to work as a system instead of the separate protocol stack, and it is termed as CLD. Energy becomes a very critical resource in WSNs, while the traditional protocol stacks are not designed for WSNs, hence new energy-efficient protocols are needed for all layers in the protocol stack [40–42,45,46]. The very first of its kind of vision for cross-layer architectures for wireless networks was proposed in [45]. In the author's work, he shows how a cross-layer design can maintain the abstraction of each layer, which supports the modularity of the architecture. Further, several attempts have taken to provide the solution for specific matrix of the wireless network in [47–51] to help improve the network as a whole. In [52], CLPs have been presented for the multihop wireless network for joint rate control and centralized scheduling in MAC. In [47], CLPs for jointly rate control and distributed scheduling have been given. Moreover, CLP congestion control, routing and MAC scheduling for network stability have been given in [48] for WN. In [49], transport and physical layer balancing through CL design for a given WN further leads to congestion control of MAC and power control of the physical layer. In [50], there is joint consideration for congestion control and a joint approach for rate control with end-to-end fairness [51]. Moreover, the amount of data and nature of the nodes that imply significance change in configuration. That motivation leads to develop some of CLD for multimedia data rate [53,54], mobile wireless network [55], third generation [56] and fourth generation wireless network [57].

A formulation for slotted synchronous TDMA MAC scheme [2], optimization of the energy consumption per bit has been proposed for interference-free WSNs. The solution presented is optimizing the commodity flow for single

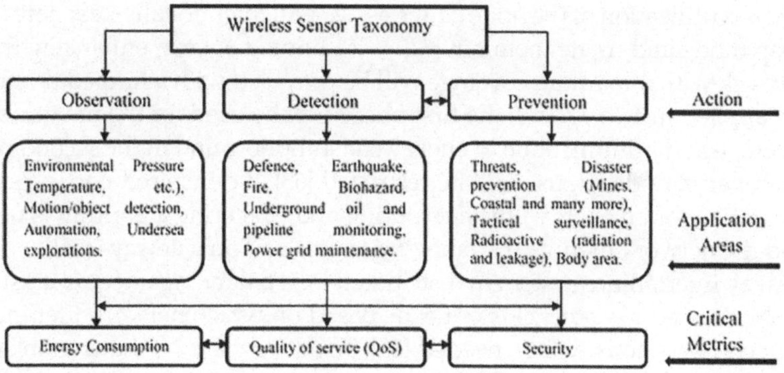

FIGURE 2.3
Action, application and metrics-based wireless classification. (From D. K. Sah, and T. Amgoth, *Computer Science Review*, Vol. 27, 112–134, 2018.)

variables, which could further extend for multicommodity in the instance when more than one sink is available to receive the data. In Figure 2.3 the classification has been made about the action and application areas, followed by critical metrics. This classification is also necessary here to make the point that, in some application scenarios, the performance matrices are definite. Instead of employing conventional architecture, the CLD has become most efficient due to its application-oriented properties.

2.2.4 Optimization Solutions for Conventional Protocol Architecture

The major issue in wireless sensor networks is effective data collection. Traditionally, data collection is based on a layered approach, which cannot provide an optimized solution. The solution to this is a novel, flow-based layer selection algorithm, which estimates the flow of data in an individual layer that will be identified and channel allocation been issued [58]. WSNs consist of multiple nodes; in the process of data aggregation, assigning random names to nodes will not be so effective if the number of nodes is high. To overcome such difficulty, a method of assigning names based on attributes and application dependent names will better identify the nodes [59]. The significant threat in wireless sensor networks is the sinkhole problem, which is caused by heavy data transfer traffic on a base station near the nodes. The failure of nodes near the base station will terminate the entire network. To overcome this defect, we propose that a division of clusters far away from the base station will contain a higher number of nodes unequally, whereas clusters close to the base station will contain fewer nodes [60]. Node sensing capabilities should be fixed with the traditional approach, which leads to a loss of energy very quickly. If we implement the process of application-dependent, sensing capabilities, we also depend on the available energy level sensing which leads to the increase of network lifetime [61]. Energy conservation is a challenging aspect of the WSNs. For the optimized utilization of the node batteries, we will intentionally keep few nodes in sleep mode and rotate them out [62]. With limited energy, enhancing the life of network with maximum coverage will be provided with a limited number of nodes applied in two phases; the first phase will be used for the maximization of coverage and minimization of energy consumption, and in the second phase, the number of nodes in use will be restricted [63]. Bio-inspired particle swarm optimization for an energy-efficient routing and clustering algorithm is used to extend the network lifetime by removing traffic load on gateways [64].

A fuzzy interference system in distributed load balancing unequal clustering in WSNs is used to detect cluster heads based on which network lifetime and energy conservations will be needed [65]. To address the NP-hard problem for area coverage, nature-based algorithms (such as improved cuckoo search and the chaotic flower pollination optimization algorithm) are used [15]. A single node is responsible for transmission of data from an inactive network. The identification of an active node with an optimized technique design of specialized matrix is proposed in [66]. In the equal distribution of nodes, there is a large scope for termination of the network. As an alternative to a traditional approach, a heterogeneous clustering of networks is used by considering the properties' adaptive, distributed, energy-aware fault-tolerant system [67]. The performance of a network is influenced by two significant factors, as the lifetime should be maximum and the delay in transmission of data which should be minimum. A novel data gathering protocol is called broadcasting, which is combined with a multi-NACK/ACK (BCMN/A) protocol and applied in intra- and interclustering [68]. Energy consumption in wireless sensor networks are highly dependent upon clustering and the identification of cluster heads, which we can achieve by a hybrid clustering algorithm based on K-means clustering and PSO; named KPSO to achieve efficient energy management of WSNs [69].

In Chapter 2, we want to cover a fascinating proposal given in [2] for a slotted synchronous TDMA MAC scheme, for minimizing the total energy of small-scale interference free WSNs proposed. It is given as a single commodity flow (ASCF) whose can be further extended to a multicommodity, where there will be more than one sink to receive the data.

The objective of the work can be summarized as follows:

a. It provides a convex optimization framework to minimize the total energy consumption, which obeys practical M-ary quadrature amplitude modulation (MQAM) at the physical layer.
b. The trade-off curve has provided the relationship between energy consumption of the network with variable delay.
c. The relationship among energy and delay is very useful in the minimization of the latency and packet delay. Therefore, a scheduling algorithm has been proposed.

ASCF–based optimization problems can reduce the convex problem because of its low-complexity solution availability. As a result, significant energy can be saved when link adaptation technique is used. To reduce possible delay, it has to minimize the worst-case packet delay. An observation has been drawn that if an outgoing link schedule is used prior to the incoming link in tree topology, then the worst-case delay any packet can experience for K hop will be $K*T$ where T is slot length of the link. It is the simple but efficient way of reducing the queening delay experiencing the packet in tree topology, shown in Figure 2.4.

If T value is different for different link then in that case KT will be as $\sum_{i=0}^{K} T^i$. Further, this algorithm extends to the complex network. It is a kind of bottom-up approach routing in which the bottom nodes in tree which is sensor nodes forward data upwards where the sink node is present. Generally in tree topology, the topmost node is sink node. The general formulation for optimizing the different layers to minimize power consumption has been explained in Equations 2.5–2.8. Other application-specific objectives are described in Figure 2.5.

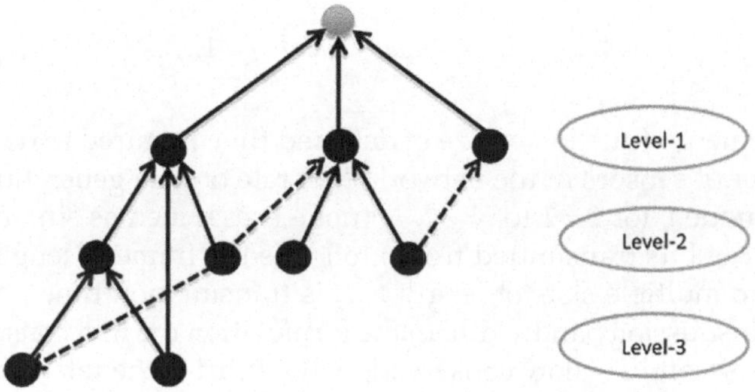

FIGURE 2.4
Data forwarding from top to bottom approach in tree topology. (From D. K. Sah, and T. Amgoth, *Computer Science Review*, Vol. 27, 112–134, 2018.)

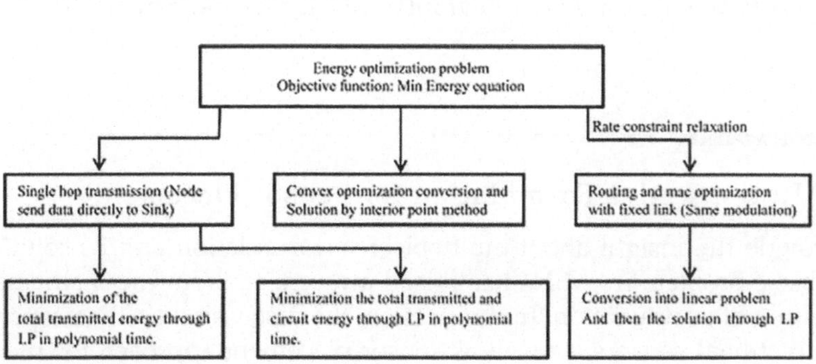

FIGURE 2.5
Flow model of optimization approach. (From D. K. Sah, and T. Amgoth, *Computer Science Review*, Vol. 27, 112–134, 2018.)

$$\text{Minimize} \sum_{i=1}^{N-1} \sum_{j=1}^{N} E(W_{ij}, t_{ij}) \tag{2.5}$$

$$\text{subject to,} \sum_{i=1}^{N-1} \sum_{j=1}^{N} t_{ij} \leq T \tag{2.6}$$

$$\sum_{j=1}^{N} (W_{ij} - W_{ji}) = R_i T, \quad i = 1,\ldots..N-1 \ \mathit{and} \ \ j = 1,\ldots..N \tag{2.7}$$

$$2Bt_{ij} \leq W_{ij} \leq BC_{ij} t_{ij} \quad , t_{ij} \in \{o, t', \ldots\ldots\ldots.\} \tag{2.8}$$

$$\forall \ i, j \in i = 1,\ldots..N-1 \ \mathit{and} \ \ j = 1,\ldots..N$$

where E is energy function of size of data and time required to transfer data, N is number of sensors in the network, R_i is rate of data generation (bits per second) at node i, for $i = 1$ to $N - 1$. N^{th} node is assumed as sink node, W_{ij} is the number of bits transmitted from i to j, T is the frame of length which is divided into multiple slots of length t', t_{ij} is transmission time on link i to j and B is transmission bandwidth. In the formulation the first constraint is for TDMA, the second for flow conservation, the third is the rate constraint for regulating the rate in between of QPSK and BPSK and the fourth is to ensure that each link has allocated an integer number for the slot. As a result, it has been concluded that joint optimization across the network, MAC and physical layers with hardware accountability are feasible and beneficial.

2.3 Multiobjective Formulation for Sensor Placement

To provide the insight about multiobjective formulation and its solution, a very basic problem in WSNs has raised awareness about the mathematical modelling of optimization. In this section, the discussion on the property of the directional sensors, as well as sensor placement approach for the rectangular region of interest (ROI) is briefly explained in the context of multiobjective optimization to provide insight to the readers about optimization significance in WSNs.

2.3.1 Directional Sensors Model and Variables

In WSN area sensing, the sensing unit is typically the omnidirectional and sensors are placed at the point *P* is able to cover the area π*R. where R is equal to the maximum line of sight (LOS_m). Here we are modelling the directional sensors and its resemblance can found in the model given for field of view in cameras.

Let us consider the 2D-model of the sensing area of sensors, which can be represented with four variables (LOC, R, LOS_m, α).

In this model, the point *'P'* covered at time instance *'t'*, if the following condition is filled:

1. dist(LOC,P) ≤ R where dist(LOC,P) is Euclidean distance between location of any sensors S_i where $i \varepsilon N$ and point *'P'* in the ROI.
2. The angle between LOS_m and line segment of *P* and LOC should in between [-α, α].

In this context, the point P is covered by directional sensor available if the length of line segment |PLOC|≤ R and angle between line segment |PLOC| to offset LOS_m is always less then abs(α).

2.3.2 Coverage Probability Estimation (CPE) of Directional Sensor

Sensor deployment in WSNs in ROI proceed in either of two ways, manual or random [70], depending upon the arrangements of ROIs. In this context, when the specific place is not easily accessible, then random deployment with the help of airplanes or another vehicle can be helpful while in a closed area (such as building block) which can be easily accessed initially for instalment. In a closed site, such as nuclear power plant or industry, the sensor that is omnidirectional is underutilised due to an obstacle such as a concrete wall and so on. Moreover, direction antennae can be cost-effective as well as used for the purpose of covering the ROI. The sensing area, or ROI, is defined as the area which is interested in sensing. Assume there are exactly no two sensors located in the same region. Besides, each point *P* is at least 1-coverage and the overlapping of the sensing range by either the top, left and the right sensing portion of sensors is supposed to be minimised. In this context, the directional sensors with offset angle α can provide sensing capability in region αR^2 and for omnidirectional the sensing region value is πR^2. The number of sensors deployed in the area is equal to *N* as mention in Table 2.2. With the help of *N*-directional sensors deployed, the probability of covering the targeted region can be given as

$$p = 1 - \left(1 - \frac{\alpha R^2}{AOI}\right)^N \tag{2.9}$$

TABLE 2.2
Notations

AOI	Area of interest or active region
LOC	Location of the sensors
R	Radius of the sensors
LOS_m	Maximum line of sight
α	Offset angle with LOS_m
S_i	Sensor with arbitrary numbering *i*
E_i	Energy of sensor *i* at initial deployment
E_{it}	Energy of sensor *i* at time instance *t*
dist(a,b)	Euclidean distance between point *a* and *b*
N	Number of sensors available
$A_{i,t}$	Area covered by sensor *i* with its top notch
$A_{i,l}$	Area covered by sensor *i* with its left sensing portion
$A_{i,r}$	Area covered by sensor *i* with its right sensing portion

For omnisensing sensors, α = π when the coverage probability once the *N* sensors is deployed is become

$$p = 1 - \left(1 - \frac{\pi R^2}{AOI}\right)^N \tag{2.10}$$

In the same way, in real-time applications, the network should prove the threshold value and the same value can used to calculate the reliability of the network. Moreover, to achieve the probability *p* (important variable for calculating reliability of the network), the number of deployed directional sensors should be at least *N* and this can simply calculate with

$$N \geq \frac{\ln(1-p)}{\ln(S - \alpha R^2) - lnAOI} \tag{2.11}$$

Again, for omnisensing sensors α = π the number of sensors required to achieve probability should be

$$N \geq \frac{\ln(1-p)}{\ln(S - \pi R^2) - lnAOI} \tag{2.12}$$

From the above equation, the ratio of directional sensors with respect to omnidirectional sensors can express in term of following:

$$\textit{Ratio of Bidrectional and omindirertional} = \frac{\ln(S - \pi R^2) - (lnAOI)}{\ln(S - \alpha R^2) - (lnAOI)} \tag{2.13}$$

2.3.3 Linear Formulation for Coverage

In this section, we will discuss that how the calculated probability estimated earlier can be accounted as the threshold value. The coverage of the area can be maximised by following the estimated probability, as well as other constraints. In Equation 2.14, the area coverage can be maximised by dividing the coverage capability of the sensors into three categories. The shaded region falls within the ROI, therefore the constraint is to make sure that request is being fulfilled.

$$maximize \sum_{i}^{N} A_{i,t} + \sum_{i}^{N} A_{i,l} + \sum_{i}^{N} A_{i,r} \tag{2.14}$$

In Equations 2.15–2.19, the required constraint is given and has explained briefly, highlighting the key issues in constraint formulation. Equation 2.15 is given to bind the sensing range of the sensors, and to not exceed from certain threshold (in our assumption: ROI). By controlling the sensing range, the depletion of energy can be reduced as sensing ranges decrease.

$$subject\ to,\ AOI \leq ROI \tag{2.15}$$

In Equation 2.16, the maximum available sensing unit is given to estimate the threshold probability. The number of sensors used to maximize the coverage should be always less than or equal to the availability of sensors.

$$N \leq N_{max} \tag{2.16}$$

Equation 2.17 is used to ensure that the point which lies inside the ROI is covered as much as possible. As we know from the study [71], in directional sensor deployment, a hole in AOI is always possible. With the help of Equation 2.17, the void can be reduced.

$$PLOC\ angle\ LOC\ \&\ LOS_m \geq |PLOC| \cos \alpha \tag{2.17}$$

In Equation 2.18, the covering angle of the directional sensors is controlled. If it increases, the depletion of the energy will be fast, therefore the covering angle is controlled by this constraint. In theory, if any of the results are showing that $\alpha = \pi$, that means that those sensors can be replaced by omnidirectional sensors for better network lifetime and energy management.

$$\alpha \leq \pi \tag{2.18}$$

In Equation 2.19, the energy is accounted so that the total time of the live network can be estimated. As previously discussed, in our two separate layers of coverage used, any of the single layer sensors will be active at time t.

$$E_i \geq E_{th} \tag{2.19}$$

2.3.4 Results and Discussion

The experimental setup for our proposed formulation are explained through the parameters of the sensors as well as ROI. The area of ROI is assumed as 10*3 $unit^2$ in which any of the three vertical blocks are considered for placement of the sensors. The building block and placement for k-coverage with orientation of the direction of the sensors can be different for all feasible solutions. If the orientation being made with Figure 2.6, then the white and black parts represent the coverage in the sensing region.

Moreover, one must understand that there can be many possible orientations and the orientation given for the angular coverage value here is varied from $\alpha = 30°$ to $\alpha = \pi$, which can be observed in Figure 2.7. The computational parameters are summarized in Table 2.3. Further, on the basic of estimated probability, coverage maximization been performed on MATLAB® version 7.10.0 with the computational capability of 2.4 GHz and RAM capacity with 4GB.

In Figure 2.7, the experiment results show the coverage rate (probability) vs. offset angle with respect to a different sensor radius ratio. It can be observed from

Figure 2.7 that as the number of sensors is increased, the probability of coverage also increases with a specific angle. The experiment can observed from Figure 2.8 with a different angle.

Moreover, as the comparison of increasing of angle from 0° to 180° with a different value of the radius of the sensor, presented in Figure 2.8. The number of sensors in this scenario is fixed based on the data set presented in Table 2.3.

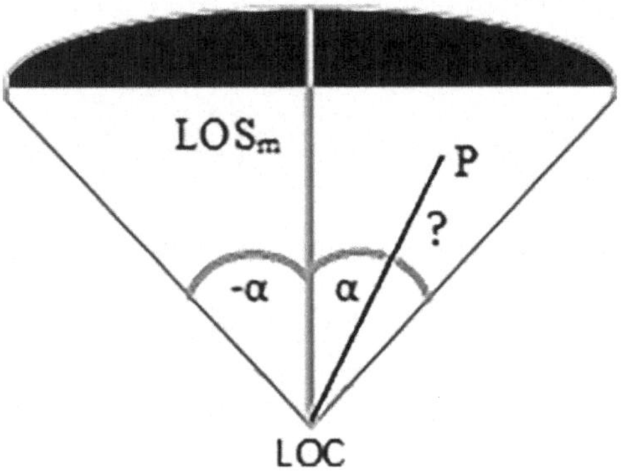

FIGURE 2.6
Directional sensor model.

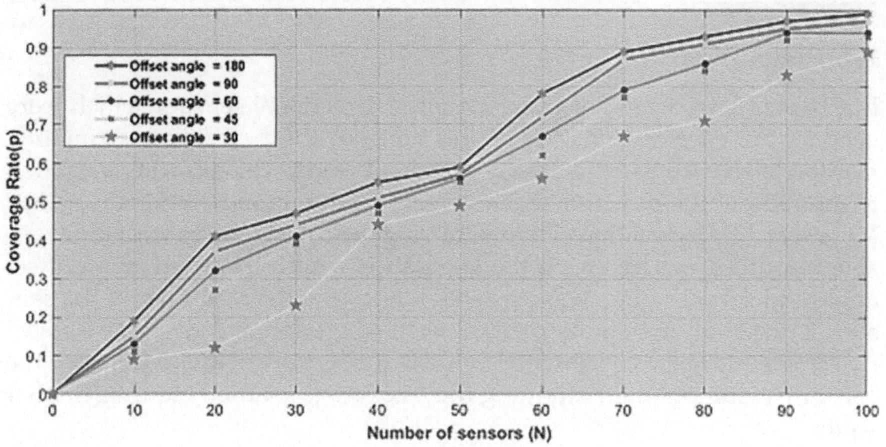

FIGURE 2.7
Effect of offset angle of sensors for threshold coverage.

TABLE 2.3
Simulation Parameters

Parameter	Default	Variation
Coverage rate *(p)*	1	0–1
Sensor number *N*	100	0–100
Offset angle α	180°	0°–180°
Sensor radius	20 unit	0–25 unit
Communication radius	40 unit	0–50 unit
Area *S*		

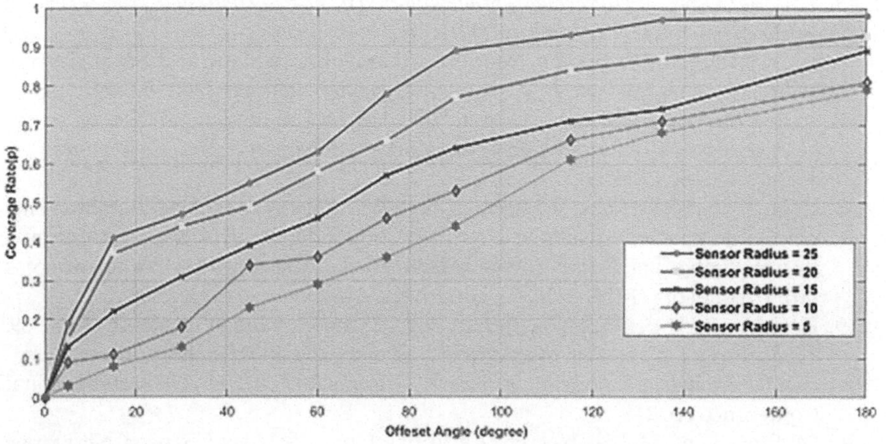

FIGURE 2.8
Effect of sensors radius on threshold coverage and on offset angle time line.

2.4 Conclusion

In Chapter 2, we were given the model of directional sensors, which were used in the coverage of ROI. In this context, the probability estimation method been used to estimate the threshold value, which further extended to the ratio of the directional sensor concerning omnidirectional sensors. Moreover, LP formulations for maximising the coverage area with the available resources was given. In the simulation result, the effect of an offset angle and sensor radius was presented. With the fixed number of the sensors, a different orientation method been suggested and the future objective is to achieve the coordinated value of the sensors in ROI. Moreover, the energy consumption for attaining the k-coverage solution requires further study.

References

1. Jennifer Yick, Biswanath Mukherjee, and Dipak Ghosal. "Wireless sensor network survey." *Computer Networks* 52(12):2292–2330, 2008.
2. Shuguang Cui, Ritesh Madan, Andrea J. Goldsmith, and Sanjay Lall. "Cross-layer energy and delay optimization in small-scale sensor networks." *IEEE Transactions on Wireless Communications* 6(10), 2007.
3. Edwin K. P. Chong and Stanislaw H. Zak. *An Introduction to Optimization, volume 76*. John Wiley & Sons, 2013.
4. A Ravindran, Gintaras Victor Reklaitis, and Kenneth Martin Ragsdell. *Engineering Optimization: Methods and Applications*. John Wiley & Sons, 2006.
5. Xin-She Yang. *Engineering Optimization: An Introduction with Metaheuristic Applications*. John Wiley & Sons, 2010.
6. Ian F. Akyildiz, Weilian Su, Yogesh Sankarasubramaniam, and Erdal Cayirci. "Wireless sensor networks: A survey." *Computer Networks* 38(4):393–422, 2002.
7. Rajeev Ranjan and Shirshu Varma. "Challenges and implementation on cross-layer design for wireless sensor networks." *Wireless Personal Communications* 86(2):1037–1060, 2016.
8. Gogu, A., Nace, D., Dilo, A., Meratnia, N., 2012. "Review of optimization problems in wireless sensor networks." J. Hamilton Ortiz (Ed.), *Telecommunications Networks: Current Status and Future Trends* InTech, Rijeka, Croatia. 153–180, openAccess http://dx.doi.org/10.5772/38360. URL ⟨http://doc.utwente.nl/80257/⟩.
9. Fischione, C., Speranzon, A., Johansson, K., Sangiovanni-Vincentelli, A., 2009. "Peer-to-peer estimation over wireless sensor networks via lipschitz optimization." *Information Processing in Sensor Networks* 2009. IPSN 2009. International Conference on. 241–252.
10. "A Survey of Multi-Objective Optimization in Wireless Sensor Networks: Metrics, Algorithms, and Open Problems." Zesong Fei, Senior Member, Fischione.

11. Fischione, C., Jonsson, U., 2011. "Fast-lipschitz optimization with wireless sensor networks applications." Information Processing in Sensor Networks (IPSN), 2011. *Proceedings of the 10th International Conference* 378–389.
12. Tan, M., Chin, Y., Tham, H., Teo, K., 2011. "Genetic algorithm based pid optimization in batch process control." Computer Applications and Industrial Electronics (ICCAIE), 2011 IEEE International Conference 162–167. http://dx.doi.org/10.1109/ICCAIE.2011.6162124.
13. Wong, C. H., Siew, Z. W., Tan, M. K., Chin, R., Teo, K., 2012. "Optimization of distributed and collaborative beamforming in wireless sensor networks." Computational Intelligence, Communication Systems and Networks (CICSyN), 2012. *Proceedings of the Fourth International Conference* 84–89. http://dx.doi.org/10.1109/CICSyN.2012.26.
14. Jabbar, S., Iram, R., Minhas, A. A., Shafi, I., Khalid, S., Ahmad, M., 2013. "Intelligent optimization of wireless sensor networks through bio-inspired computing: Survey and future directions." *Int. J. Distrib. Sens. Netw.* 2013, 13. http://dx.doi.org/10.1155/2013/421084, (URL ⟨http://downloads.hindawi.com/journals/ijdsn/2013/421084.pdf⟩).
15. Binh, Huynh Thi Thanh, Nguyen Thi Hanh, and Nilanjan Dey. "Improved cuckoo search and chaotic flower pollination optimization algorithm for maximizing area coverage in wireless sensor networks." *Neural Computing and Applications* (2016): 1–13.
16. Dey, Nilanjan et al. "Wireless Capsule Gastrointestinal Endoscopy: Direction-of-Arrival Estimation Based Localization Survey." *IEEE Reviews in Biomedical Engineering* 10 (2017): 2–11.
17. Kulkarni, R., Venayagamoorthy, G., 2011. "Particle swarm optimization in wireless-sensor networks: A brief survey." *IEEE Trans. Syst., Man, Cybern., Part C: Appl. Rev.* 41 (2), 262–267. http://dx.doi.org/10.1109/TSMCC.2010.2054080.
18. Kennedy, J., Eberhart, R., 1997. "A discrete binary version of the particle swarm algorithm." *Systems, Man, and Cybernetics, 1997. Computational Cybernetics and Simulation.* 1997 IEEE International Conference Vol. 5. 4104–4108. http://dx.doi.org/10.1109/ICSMC.1997.637339.
19. Z. Yun, X. Bai, D. Xuan, T. H. Lai, and W. Jia, "Optimal deployment patterns for full coverage and k- connectivity ($k \leq 6$) wireless sensor networks." *IEEE/ACM Trans. Netw.* Vol. 18, No. 3, 934–947, Jun. 2010.
20. K. S. S. Rani and N. Devarajan, "Multiobjective sensor node deployment in wireless sensor networks." *Int. J. Eng. Sci.* Vol. 4, No. 4, 1262–1266, Apr. 2012.
21. M. Abo-Zahhad, N. Sabor, S. Sasaki, and S. M. Ahmed, "A centralized immune-Voronoi deployment algorithm for coverage maximization and energy conservation in mobile wireless sensor networks." *Inf. Fusion* Vol. 30, 36–51, Jul. 2016.
22. K. Sakai, M.-T. Sun, W.-S. Ku, T. H. Lai, and A. V. Vasilakos, "A framework for the optimal k-coverage deployment patterns of wireless sensors." *IEEE Sensors J.* Vol. 15, No. 12, pp. 7273–7283, Dec. 2015.
23. A. Konstantinidis, K. Yang, Q. Zhang, and D. Zeinalipour-Yazti. "A multi-objective evolutionary algorithm for the deployment and power assignment problem in wireless sensor networks." *Comput. Netw.* Vol. 54, No. 6, 960–976, Apr. 2010.
24. J. Chen, S. He, Y. Sun, P. Thulasiraman, and X. Shen, "Optimal flow control for utility-lifetime tradeoff in wireless sensor networks." *Comput. Netw.* Vol. 53, No. 18, 3031–3041, Dec. 2009.

25. J. H. Zhu, K.-L. Hung, B. Bensaou, and F. Nait-Abdesselam, "Ratelifetime trade-off for reliable communication in wireless sensor networks." *Comput. Netw.* Vol. 52, No. 1, 25–43, Jan. 2008.
26. Y. Zhao, J. Wu, F. Li, and S. Lu, "On maximizing the lifetime of wireless sensor networks using virtual backbone scheduling." *IEEE Trans. Parallel Distrib. Syst.* Vol. 23, No. 8, 1528–1535, Aug. 2012.
27. V. Shah-Mansouri and V. W. S. Wong, "Lifetime-resource tradeoff for multicast traffic in wireless sensor networks." *IEEE Trans. Wireless Commun.* Vol. 9, No. 6, 1924–1934, Jun. 2010.
28. R. R. Rout and S. K. Ghosh, "Enhancement of lifetime using duty cycle and network coding in wireless sensor networks." *IEEE Trans. Wireless Commun.* Vol. 12, No. 2, 656–667, Feb. 2013.
29. C.-F. Wang, J.-D. Shih, B.-H. Pan, and T.-Y. Wu, "A network lifetime enhancement method for sink relocation and its analysis in wireless sensor networks." *IEEE Sensors J.* Vol. 14, No. 6, 1932–1943, Jun. 2014.
30. M. Elhoseny et al., "Balancing energy consumption in heterogeneous wireless sensor networks using genetic algorithm." *IEEE Commun. Lett.* Vol. 19, No. 12, 2194–2197, Dec. 2015.
31. Y. Yao, Q. Cao, and A. V. Vasilakos, "EDAL: An energy-efficient, delay-aware, and lifetime-balancing data collection protocol for heterogeneous wireless sensor networks." *IEEE/ACM Trans. Netw.* Vol. 23, No. 3, 810–823, Jun. 2015.
32. C.-T. Cheng and H. Leung, "A multi-objective optimization framework for cluster-based wireless sensor networks." *Proc. IEEE Int. Conf. Cyber Enabled Distrib. Comput. Knowl. Disc. (CyberC)*, Sanya, China, Oct. 2012, 341–347.
33. H. Li, C. Wu, D. Yu, Q.-S. Hua, and F. C. M. Lau, "Aggregation latency-energy tradeoff in wireless sensor networks with successive interference cancellation." *IEEE Trans. Parallel Distrib. Syst.* Vol. 24, No. 11, 2160–2170, Nov. 2013.
34. J. Lu, X. Wang, L. Zhang, and X. Zhao, "Fuzzy random multi-objective optimization based routing for wireless sensor networks," *Soft Comput.* Vol. 18, No. 5, 981–994, May 2014.
35. Q. Gao, Y. Zuo, J. Zhang, and X.-H. Peng, "Improving energy efficiency in a wireless sensor network by combining cooperative MIMO with data aggregation." *IEEE Trans. Veh. Technol.* Vol. 59, No. 8, 3956–3965, Oct. 2010.
36. W. Fang, F. Liu, F. Yang, L. Shu, and S. Nishio, "Energy-efficient cooperative communication for data transmission in wireless sensor networks." *IEEE Trans. Consum. Electron.* Vol. 56, No. 4, 2185–2192, Nov. 2010.
37. J. M. Lanza-Gutierrez and J. A. Gomez-Pulido, "Assuming multiobjective metaheuristics to solve a three-objective optimization problem for relay node deployment in wireless sensor networks." *Appl. Soft Comput.* Vol. 30, 675–687, May 2015.
38. A. Sahoo and C. Shanti, "DGRAM: A delay guaranteed routing and MAC protocol for wireless sensor networks." *IEEE Trans. Mobile Comput.* Vol. 9, No. 10, 1407–1423, Oct. 2010.
39. C.-T. Cheng, C. K. Tse, and F. C. M. Lau, "A delay-aware data collection network structure for wireless sensor networks." *IEEE Sensors J.* Vol. 11, No. 3, 699–710, Mar. 2010.
40. Stallings, William. *Wireless Communications and Networks.* Pearson Education India, 2009.
41. Yick, Jennifer, Biswanath Mukherjee, and Dipak Ghosal. "Wireless sensor network survey." Computer Networks 52.12 (2008): 2292–2330.

42. Akyildiz, Ian F. et al. "Wireless sensor networks: A survey." *Computer networks* 38.4 (2002): 393–422.
43. Stallings, William. *Handbook of Computer Communications Standards; Vol. 1: The Open Systems Interconnection (OSI) Model and OSI-Related Standards.* Macmillan Publishing Co., Inc., 1987.
44. Al-Karaki, Jamal N., and Ahmed E. Kamal. "Routing techniques in wireless sensor networks: A survey." *IEEE Wireless Communications* 11.6 (2004): 6–28.
45. Shakkottai, Sanjay, Theodore S. Rappaport, and Peter C. Karlsson. "Cross-layer design for wireless networks." *IEEE Communications Magazine* 41.10 (2003): 74–80.
46. B. Fu, Y. Xiao, H. J. Deng and H. Zeng, "A Survey of Cross-Layer Designs in Wireless Networks." *IEEE Communications Surveys and Tutorials* Vol. 16, No. 1, 110–126, First Quarter 2014. doi: 10.1109/SURV.2013.081313.00231.
47. L. Bui, A. Eryilmaz, R. Srikant, and X. Wu, "Joint congestion control and distributed scheduling in multihop wireless networks with a nodeexclusive interference model." *Proc. IEEE INFOCOM* 2006.
48. A. Eryilmaz and R. Srikant, "Joint congestion control, routing and MAC for stability and fairness in wireless networks." *IEEE J. Sel. Areas Commun.* Vol. 24, No. 8, 1514–1524, Aug. 2006.
49. M. Chiang, "Balancing transport and physical layer in multihop wireless networks: Jointly optimal congestion and power control." *IEEE J. Sel. Areas Commun.* Vol. 23, No. 1, 104–116, Jan. 2005.
50. L. Chen, S. H. Low, and J. C. Doyle, "Joint congestion control and media access control design for wireless ad hoc networks." *Proc. IEEE INFOCOM* Miami, FL, Mar. 2005, 2212–2222.
51. X. Wang and K. Kar, "Cross-layer rate control for end-to-end proportional fairness in wireless networks with random access." *Proc. ACM Mobihoc* Urbana-Champaign, IL, May 2005, 157–168.
52. X. Lin and N. B. Shroff, "Joint rate control and scheduling in multihop wireless networks." *Proc. IEEE Conf. Decision and Control,* Paradise Island, Bahamas, Dec. 2004, 1484–1489.
53. E. Setton, T. Yoo, X. Zhu, A. Goldsmith, and B. Girod, "Cross-Layer Design of Ad Hoc Networks for Real-Time Video Streaming." *IEEE Wireless Communications* August 2005.
54. Khan, Shoaib et al. "Application-driven cross-layer optimization for video streaming over wireless networks." *IEEE Communications Magazine* 44.1 (2006): 122–130.
55. F. Foukalas, V. Gazis, and N. Alonistioti, "Cross-Layer Design Proposals for Wireless Mobile Networks: A Survey and Taxonomy." *IEEE Commun. Surveys and Tutorials* 70–85, 2008.
56. H. Jiang, W. Zhuang, and X. Shen, "Cross-Layer Design for Resource Allocation in 3G Wireless Networks and Beyond." *IEEE Commun. Mag.* Dec. 2005.
57. G. Carneiro, J. Ruela, and M. Ricardo, "Cross-Layer Design in 4G Wireless Terminals." *IEEE Wireless Commun.* Vol. 11, No. 2, 7–13, April 2004.
58. Navaz, A. S. Syed, and Dr G. M. Kadhar Nawaz. "Flow Based Layer Selection Algorithm for Data Collection in Tree Structure Wireless Sensor Networks." *International Journal of Applied Engineering Research* 11.5 (2016): 3359–3363.
59. Rajesh, M., and J. M. Gnanasekar. "Constructing Well-Organized Wireless Sensor Networks with Low-Level Identification." *World Engineering & Applied Sciences Journal* 7.1 (2016).

60. Rao, P. C. Srinivasa, and Haider Banka. "Novel chemical reaction optimization based unequal clustering and routing algorithms for wireless sensor networks." *Wireless Networks* 23.3 (2017): 759–778.
61. Hu, Yanling et al. "Mobile target detection in wireless sensor networks with adjustable sensing frequency." *IEEE Systems Journal* 10.3 (2016): 1160–1171.
62. Komurlu, Caner, and Mustafa Bilgic. "Active Inference and Dynamic Gaussian Bayesian Networks for Battery Optimization in Wireless Sensor Networks." *AAAI Workshop: AI for Smart Grids and Smart Buildings*. 2016.
63. Abo-Zahhad, Mohammed et al. "A centralized immune-Voronoi deployment algorithm for coverage maximization and energy conservation in mobile wireless sensor networks." *Information Fusion* 30 (2016): 36–51.
64. Azharuddin, Md, and Prasanta K. Jana. "Particle swarm optimization for maximizing lifetime of wireless sensor networks." *Computers & Electrical Engineering* 51 (2016): 26–42.
65. Baranidharan, B., and B. Santhi. "DUCF: Distributed load balancing Unequal Clustering in wireless sensor networks using Fuzzy approach." *Applied Soft Computing* 40 (2016): 495–506.
66. Chen, Wei, and Ian J. Wassell. "Optimized node selection for compressive sleeping wireless sensor networks." *IEEE Transactions on Vehicular Technology* 65.2 (2016): 827–836.
67. Deniz, Fatih, Hakki Bagci, and Ibrahim Korpeoglu. "An adaptive, energy-aware and distributed fault-tolerant topology-control algorithm for heterogeneous wireless sensor networks." *Ad Hoc Networks* 44 (2016): 104–117.
68. Dong, Mianxiong et al. "Joint optimization of lifetime and transport delay under reliability constraint wireless sensor networks." *IEEE Transactions on Parallel and Distributed Systems* 27.1 (2016): 225–236.
69. Solaiman, Basma. "Energy optimization in wireless sensor networks using a hybrid k-means pso clustering algorithm." *Turkish Journal of Electrical Engineering & Computer Sciences* 24.4 (2016): 2679–2695.
70. R. Mulligan, and H. M. Ammari, "Coverage in wireless sensor networks: A survey." *Network Protocols and Algorithms* Vol. 2, No. 2, 27–53, 2010.
71. D. K. Sah, and T. Amgoth, "Parametric survey on crosslayer designs for wireless sensor networks." *Computer Science Review* Vol. 27, 112–134, 2018.

3

Applications of Machine Learning in Wireless Sensor Networks

Ramchandra S. Mangrulkar and Prashil D. Negandhi

CONTENTS

3.1 Introduction

The development of ML was started in the late 1950s as one of the innovative techniques for developing artificial intelligence (AI) systems. As time passed, the main focus shifted to developing algorithms which were computationally feasible, outperformed conventional methodologies, and were adaptive in nature. In the last 15 years, machine learning techniques have emerged as an impressive technology with various applications involving classification, regression and density estimation with potentially diverse applications for wireless sensor networks, Internet of Things, bioinformatics, speech recognition, spam mail detection, computer vision, computer security, fraud detection, social networks, etc. The algorithms come from many diverse fields, such as statistics, neuroscience, mathematics and computer science. Some of the reasons why ML techniques play an important role in wireless sensor applications are:

1. Wireless sensor networks monitor nonstatic environments which would need the node to rapidly adjust to new circumstances. The nodes, if required, must be able to dynamically change its locality based on the deployment of the network.
2. WSNs systems may be interested in collecting information about unreachable nodes' environment, or diagnosing nodes whose physical properties are not acceptable, such as battery level or locality.

3. WSNs are usually deployed in diverse and unpredictable environments where researchers and architects might fail to predict the behaviour of the system and hence require complex algorithms to solve it, especially for routing issues.
4. Wireless sensor nodes generate a large amount of data and it might be possible that a fair amount of it is correlated (data collected from neighboring nodes could vary negligibly). In such a scenario, transferring all the nodes' data isn't feasible. Thus algorithms are needed to decide which data should be sent to preserve the battery life of the nodes.
5. Recent challenging applications of WSNs are machine-to-machine (M2M) communications, cyber-physical systems (CPS) and the Internet of Things (IoT). Data analytics have been introduced with promising results to help design and develop intelligent, self-oriented, decision-making control systems for WSNs.

However, there are a few major limitations that should be considered while designing systems using machine learning techniques in wireless sensor networks. A few of them are:

1. WSN is a limited resource framework. Predicting accurate mapping functions and extract relationships among generated data samples drains a lot of the node's energy.
2. Learning by example requires a large dataset of samples to achieve the desired generalization capabilities and accuracy. Moreover, the designer fails to get full control over the knowledge formulation process.

3.2 Machine Learning Algorithms and Applications in WSN

Machine learning is a branch of artificial intelligence (AI) which deals with providing systems with the ability to dynamically learn and improve their models from experience (without being explicitly programmed to do so). ML algorithms focus on the development of systems that can access data and use it to perfect their predictions as more data becomes readily available. For the learning process, past observations or historical data are required in order to look for patterns and better make decisions in the future based on the trends identified. ML techniques have found diverse applications in many sectors, including healthcare, finance, automation, etc. Recently, attempts are being made to leverage these algorithms to the field of networking to further enhance the capabilities of computer networks.

WSNs could potentially benefit a lot from ML algorithms. This is primarily because the individual sensor nodes are usually deployed in an unknown environment that needs to be mapped or observed. This environment could be both an outdoor environment or an indoor environment. A recent implementation of using WSNs for a home environment has been explained in [1]. In this system, deploying sensors for monitoring the health of home owners was proposed. Due to the uncertainty of these dynamic surroundings, it becomes hard for programmers to accurately program conventional systems that would power these networks. Thus, ML techniques are proposed so that sensor nodes can adapt and operate efficiently in such circumstances. Researchers have adopted various ML algorithms (explained in short) and used them in various issues concerning WSNs like routing, clustering, data aggregation, security, localization, data integrity, quality of service, etc. The various ML approaches and their associated applications for WSNs are explained as follows:

A. *Supervised learning* is defined as the task of inferring a model based on labelled training data. This model represents the learned relation between the labelled inputs and outputs. Supervised learning algorithms have found diverse applications in WSNs. Some common approaches with their scope are as follows:

- *K-nearest neighbors* (KNN) is a one of the simplest ML algorithms. It stores all the available cases with their respective classifications. New cases are classified based on a similarity measure (e.g. distance functions). This classification is done by a majority vote of its immediate neighbors and the case is assigned to the class most frequent amongst its K nearest neighbors (thus the name) measured by the similarity measure. If there is just one neighbor (K = 1), then the case is simply assigned to the class of this nearest neighbor.

  KNN's application for the case of query processing in WSNs has been studied by the researchers in [2]. They have proposed a modified KNN called the Parallel Concentric-circle Itinerary-based KNN (PCIKNN) query processing technique. PCIKNN works by deriving different itineraries based on optimization of either query latency or energy consumption. It operates in three phases: the routing phase, the KNN boundary estimation phase and the query dissemination phase. To further increase the accuracy of the KNN query, they have proposed the use of dynamically refining the estimated KNN boundary while the query is propagating within it. Simulations were performed on CSIM [3] for 1000 nodes randomly distributed in a 500×500 m^2 region. Results for three itinerary-based KNN algorithms (i.e. PCIKNN, IKNN and DIKNN [4]) were compared with parameters like query latency and accuracy and energy consumption in mind. To prevent bias, environmental factors such as the number of sample sizes (i.e. K for KNN queries), the network density, the node mobility and the failure rate of nodes were modified to obtain fair results. It was concluded by the researchers that PCIKNN outperforms these contemporary approaches in all the mentioned parameters. Interested readers for the application of KNN in query processing for WSNs should also refer [5] and [6]. KNN has also been successfully implemented for detecting abnormalities in spatial distributions of data. Further reading on the same can be found in [7]. An overview of itinerary-based KNN query processing, as depicted by the researchers of [2], is given in Figure 3.1.

- A *decision tree* (DT) can best be described as a decision support tool. It does this by using a tree-like graph or model of decisions and their possible consequences (including chance event outcomes, resource costs and utility). It can be explained succinctly as an algorithm that just consists of conditional control systems. It is of primary consideration to extend the lifetimes of all WSNs, as each sensor has limited computation, energy and memory resources. Appropriate cluster-head node elections that decrease the long-distance communication between sensors and sink nodes would reduce energy consumption and thus enhance the network lifetime.

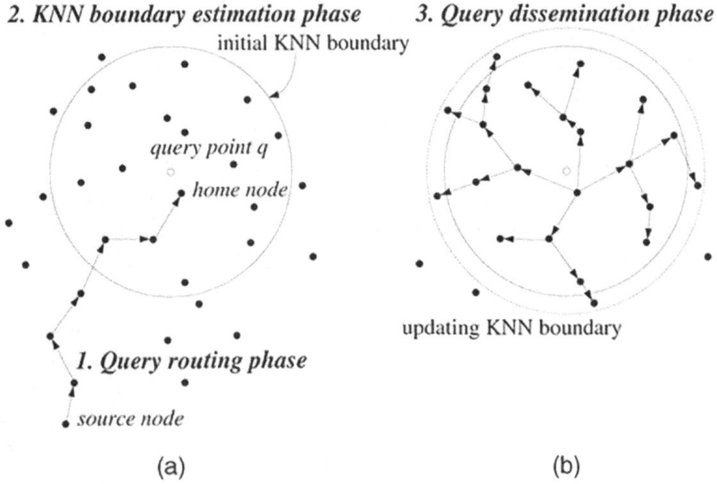

FIGURE 3.1
Itinerary-based KNN query processing.

The researchers in [8] have studied the use of the DT algorithm to select the ideal node as the cluster-head. The proposed scheme is based on four factors: the node's distance from the cluster centroid, the node's battery reserve, the node's mobility and its vulnerability index. Nodes that have high energy, low mobility, low vulnerability and are more close to the cluster centroid are more likely to be declared cluster-heads. Simulations were run by the researchers and it was found the proposed scheme produced well distributed clusters and thus helped in prolonging the network lifetime. A simplified diagram of the communication using cluster-heads is given in Figure 3.2.

Another key factor that must be taken into consideration while deploying WSNs is the channel that the nodes would communicate on. WSNs that have a dynamic channel selection that would fare much better in conditions where the electromagnetic spectrum is crowded with other existing networks. Since WSNs

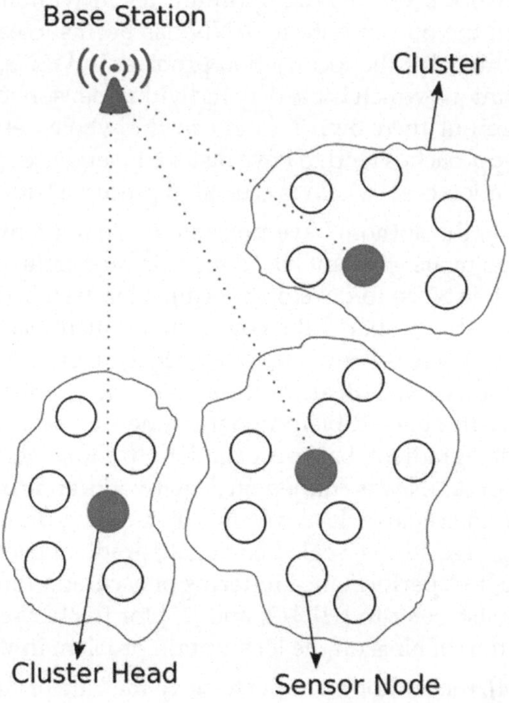

FIGURE 3.2
Communication using cluster-head.

primarily work on the ISM (industrial, scientific and medical) band, using the proper channel to communicate gains would have even more significance if the deployment is in urban areas. Therefore, intelligent methods like cognitive radio (CR) are used in the case of WSNs. The authors of [9] have used the supervised ML technique of DT for channel selection. They built a channel predictor that takes the inputs of RSSI (received signal strength indicator), the number of transmission attempts, reasons for each failed attempt and performance data from the last received packet. Using these parameters, the predictor outputs the quality of the link by estimating the number of transmission attempts for the next N packets (the prediction window). The authors have compared various algorithms and formulated their findings in a table. It was found overall that the ML models showed improved channel selection and increased the network lifetime.

- *Neural networks* (NN) are Also called as *artificial neural networks* (ANN) and mimic the way the human brain works. They attempt to identify the underlying relationships in a datasets (both labelled and unlabelled). ANN are extremely useful because of their ability to adapt to changing input. This ensures that the network produces the optimum result without the need to redesign the output criteria. ANN has been successfully implemented to solve the localisation problem in WSNs. Localisation is a technique which is used by individual sensor nodes in WSNs to figure out their own location in the whole network. ANN–based approaches tend to have faster convergence speed and low computation cost (which is crucial to preserve battery life).

In [10], the authors have presented a comparative analysis of some conjugate gradient based feedforward artificial neural networks (FFANNs) to develop a localisation framework in WSNs. The analysis compares the conjugate gradient backpropagation with the Fletcher–Reeves update, conjugate gradient backpropagation with Powell–Beale restarts, conjugate gradient backpropagation with Polak–Ribière updates, one-step secant, and scaled conjugate gradient backpropagation training algorithms. The model for ANN was trained on 30 networks and run in a perfect simulation scenario. It was found that among the five conjugate training algorithms, scaled conjugate gradient backpropagation had the best performance in terms of accuracy. Interested readers can also refer to [11], [12] and [13] for further reading on the application of NNs on the localisation problem in WSNs.

In [14], the author attempts to solve the barrier coverage problem for randomly scattered sensor nodes in WSNs. The barrier coverage problem in WSNs is defined as the ability to set-up an efficient barrier line post node deployment. In randomly deployed networks, there might not be a full end-to-end barrier line and thus these breaks must be filled in using extra nodes. The author uses ANN in predicting the number of additional static nodes or mobile nodes that must be deployed to set up the barrier line. He has attempted to design the overall network in such a way that there is a high degree of cost–benefit ratio between the static nodes and the added mobile ones. Results show that ANN can be successfully used with a high degree of predictability (R-factor above 0.99 on test data). This algorithm can further be extended to the prediction of k-barrier creation in various deployment conditions and also to the problem of the mobile nodes' movement estimation. This is because the system's complexity remains similar.

The versatility of NNs have even led to them being applied even to the case of clustering [15], event detection [16], security [17] and data integrity [18] in modified forms. Interested readers review the aforementioned literature for a detailed explanation of the same. A simplified diagram given by the authors of [14] is shown in Figure 3.3.

- *Support vector machines* (SVMs) are a classification of algorithms that work by finding the hyperplane to maximize the margin between the two classes. The support vectors are the vectors (cases) that define the hyperplane. An SVM model is represented by examples as points in space. The model is calculated such that the examples of discrete classes are divided by a gap that is as wide as possible. New examples are then classified by mapping

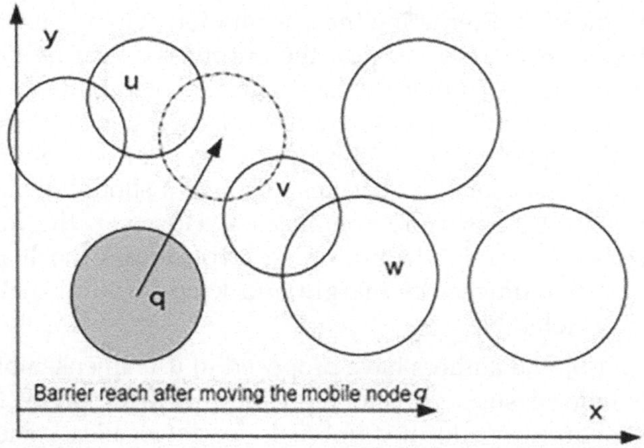

FIGURE 3.3
Effect of mobility.

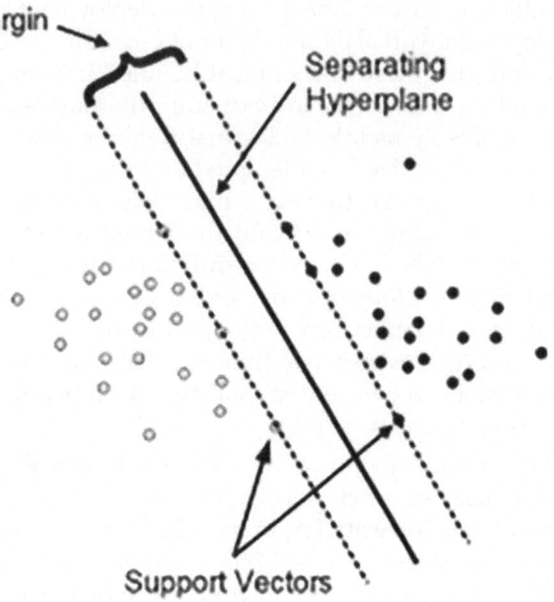

FIGURE 3.4
Support vector and hyperplane.

them into that same space and calculating which side of the separating hyperplane they lie on, as shown in Figure 3.4.

Support vector machines are also used as a regression method. Support vector regression (SVR) maintains all the key features of SVM (e.g. maximum margin). SVR thus uses the same principles, with only a few minor differences, as SVM for classification. Predicting the information at hand becomes very difficult, primarily because the output is a real number. This is because real numbers have infinite possibilities. Thus for regression, a margin of tolerance called epsilon is set in approximation to the SVM which would already have been requested from the problem. But besides this fact, it should be noted that the algorithm is more complicated. However, the main idea remains the same: to minimize error, locate the hyperplane which maximizes the margin and keep in mind that a small error is tolerable.

In [19], the authors have proposed to use an ensemble implementation of support vector regression (SVR) for solving the problem of target localisation in WSNs. Target localisation is defined as the estimation of an unknown target and is one of the main issues in WSNs. Although large-scale scale WSN deployment has become feasible due to technological advances, issues like limited communication and dimensionality still exist. These issues significantly hamper the application of ML techniques on large-scale WSNs. To overcome these challenges, the authors have tested an ensemble combination of SVRs to determine the target location using a position detection algorithm. They have compared it with conventional SVR and concluded that the proposed methodology provides a good indicator and outperforms a conventional SVR in terms of robustness and accuracy. The testing was carried out in both indoor and outdoor settings on a modest number of sensor nodes. However the authors have posited that their technique can be even more effective on larger networks to solve the problem of communication loss and explosive amounts of data. SVMs have also found application in solving the localisation problem in WSNs [20–21]. Interested readers can survey the literature to learn more about it.

- *Logistic and linear regression*: In statistical modelling, regression analysis is defined as a set of statistical processes that estimate the relationship amongst it's variables. There are many approaches for modelling and analysing several variables. These techniques focus on figuring out the relationship between a dependent variable and one or more independent variables termed as predictors. Thus, regression analysis helps one understand how the typical value of the dependent variable (or 'criterion variable') changes when any one of the independent variables is varied (the other independent variables should be fixed).

In [22], the authors have proposed using the machine learning technique linear regression for WSN data reduction. They have proposed a methodology that boosts the accuracy of regression by implementing multivariate linear regression. That is, instead of using just one variable (e.g. time/epoch) to predict the dependent variable (e.g. temperature) two are used. The model takes into consideration the correlation between two readings of data gathered by the sensor node and the time variable/epoch. Although simple linear regression methodologies are energy efficient, the researchers have found that they fail to achieve the desired accuracy that is then transferred to the user application. The proposed model attempts to reduce the redundant data being sent to the sink by employing the use of correlation packets if a neighboring node has the same coefficients of the multiple linear regression function (namely beta and alpha). Based on their experimentation, it was found that their model outperformed simple linear regression methodologies, although spent a little more energy in computation.

In [23], the authors have proposed an anomaly-based intrusion detection system (IDS) that uses the binary regression (BLR) statistical methodology to detect attacks in WSNs. Since WSNs are often deployed in uncontrollable environments, intruders are a very real risk to the network. Intruders might attack the network and target specific network layer vulnerabilities, however normally these attacks affect other layers of the network too. The authors have proposed a methodology to detect these attacks by monitoring and evaluating local sensor activity at multiple sensor networks. Using BLR, the IDS classifies the activity as either benign or malicious. The independent variables considered to train the BLR model were the number of packets forward, data packets received and the number of packets dropped. The attacks considered were selective forward and blackhole attacks. Running the model, the authors found that it had an accuracy range of 88–100%. An advantage of the proposed system was that it was able to correctly classify even novel attacks. WSNs are commonly used in a wide range of applications from monitoring changes in environment to preventing disasters like tsunamis and forest fires. Thus, WSNs should be able to rapidly assess changes and analyse sensor data distributively to ensure lower battery consumption. If however some sensors are not functioning properly (by physical damage or faulty measurements) then the integrity of the entire system's ability to perform is called into question.

In [24], the authors have proposed a model based on logistical regression (LR) that attempts to identify the nodes that are measuring faulty data in WSNs. The LR ML method has been divided into two steps and distributed to sink nodes for the same. In the first step, all sensor nodes send their data (input data) to the sink node, which then executes the learning step and then sends the optimized LR algorithm to the corresponding sensor nodes. Individual sensors then execute the second step which computes the status of their new measured data. If the data is found to be faulty, it isn't transmitted to conserve battery life. This algorithm, however, only finds out if the data is faulty. It might be entirely possible that due to environmental effects, even normal sensor nodes send faulty data. The approach has been summarized in Figure 3.5.

B. *Unsupervised learning* is defined as the training of an AI algorithm which takes as input information that is neither classified nor labelled. This information then allows the algorithm to perform actions without guidance. In unsupervised learning, an AI system must make its own categories by grouping this unsorted information

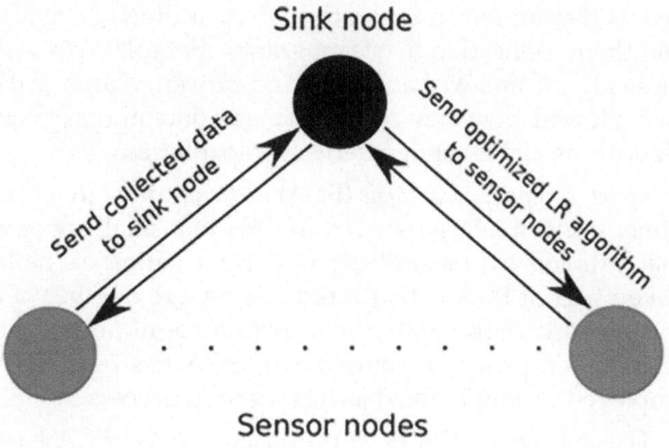

FIGURE 3.5
Sink node communication.

according to similarities and differences. AI systems that are capable of unsupervised learning are associated more often with generative models (this, however, doesn't imply they don't use a retrieval-based approach). Some of the commonly used unsupervised learning algorithms, with their application for WSNs, are as follows:

- *K-means clustering* is a very popular unsupervised learning algorithm, originally from signal processing, that is used for cluster analysis in data mining applications. It is used when the dataset is unlabelled, i.e. has no defined classes or groups. The goal of K-means clustering is to puzzle out groups in the data, with the number of groups represented by the variable K. Based on the features provided the algorithm iteratively assigns each data point to one of K groups. Clustering of data points is done using feature similarity.

  The K-means algorithm has found primary application in clustering spatially distributed data. The authors of [25] have proposed the use of distributed clustering algorithms (both deterministic and probabilistic schemes) to solve a centralised problem in a distributed manner at each sensor node. They have proposed two algorithms: distributed K-means (DKM) and distributed expectation minimization (DEM) for partitional clustering by capitalizing on a consensus-based formulation and parallel optimization tools. DKM attempts to use spatially distributed observation sets to obtain a unified clustering rule across the nodes. DEM uses local observations sets to estimate the parameters defining a mixture of pdfs that drew the observations. An interested reader should go through [25] and the proofs therein for an in depth understanding of the algorithms and their application to oceanographic data. The convergence of these algorithms was analysed and proven. Numerical analysis also showed that they outperformed contemporary centralised algorithms and extended network capabilities.

- *Principal component analysis* (PCA) is a technique that reduces the dimensions of a large set of variables to a small set of variables that still contain most of the vital information of the large set. A key part of PCA is that it reduces a large number of possibly correlated variables into a smaller number of uncorrelated variables called principal components. PCA has been successfully employed to simply the data aggregation process in WSNs.

  The authors of [26] have proposed a novel PCA based algorithm, called the Local PCA, to prolong the network lifetime by reducing the number of transmitted bytes in the intermediate nodes. The algorithm works by aggregating the incoming packets of the intermediate node into one packet and thus the node has to send out just one packet. A key differentiating factor of the proposed model is that Local PCA allows the base station to access the observations and it computes the reconstruction error. This allows the base station to monitor the accuracy of the algorithm and thus dynamically tune its update rate to perform better. Simulations were run by the researchers and it was found that the proposed model showed superior performance for the accuracy and efficiency metrics. A simplified diagram as given by the researchers of [26] is given in Figure 3.6.

  PCA is a powerful technique with potentially diverse applications to WSNs. However, a centralised implementation of the

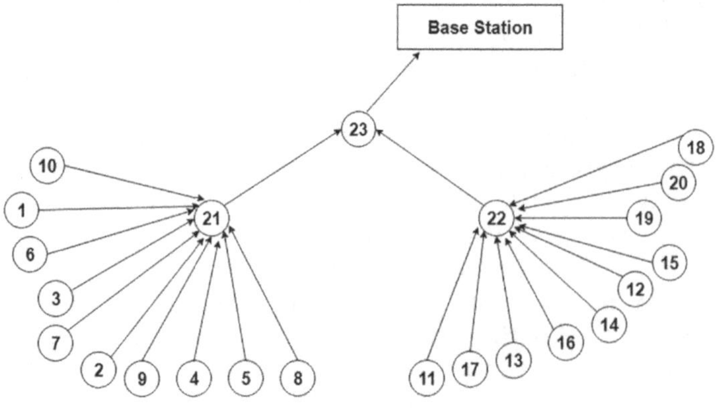

FIGURE 3.6
Local PCA model.

same is inefficient in terms of fault tolerance, scalability and energy consumption. Thus, the authors of [27] have proposed fully-distributed, consensus-based algorithms that are guaranteed to converge to the global results. They use just the local communications amongst neighbors, regardless of the distribution of data or the network sparsity. They are consensus-based PCA (CBPCA), which works by calculating the eigenvectors of local covariance matrices, and CB-EM-DPCA, a distributed version of the expectation maximization algorithm. Both these algorithms are robust (based on simulations) and provide a flexible trade-off between the desired performance (tightness of agreement on the local values with the global solution of PC) and the resulting energy expenditure (number of consensus iterations). Interested readers can also look at [28] for an in depth understanding of using PCA in combination with compressive sensing to solve the data acquisition problem for the case of WSNs.

C. *Reinforcement learning* (RL) is defined (in the context of ML) as a type of dynamic programming that trains algorithms using a system of reward and punishment. Rewards are awarded for performing an action correctly and penalties for performing incorrectly. This method is highly inspired by behaviourist psychology and is similar to how a child learns to perform a new task. The most widely known RL algorithm is the Q-learning algorithm [29]. Q-learning, as shown in Figure 3.7 is a model-free RL technique. Moreover, it can be used to select optimal actions for any given (finite) Markov decision process (MDP). The algorithm works by learning an action-value function which is denoted by Q(s, a). The ultimate goal of the algorithm is to give the expected utility of taking a given action a, in a given state s, and continue to follow the optimal policy thereafter. A policy is a rule that the agent uses to select actions based on its current state. When such an action-value function is learned, the optimal policy can be constructed by simply selecting the action which has the highest value in each state.

In WSN application, there exists a multitude of tasks which require computation at each sensor node. However, due to energy limitations and limited computing abilities of each sensor node, effective scheduling of these tasks will have a strong impact on performance and energy consumption. In [30], the authors have proposed the use of cooperative reinforcement learning for scheduling these tasks. The authors have proposed a model that trades the application performance and minimum energy consumption using a weighted reward function. Thus, it can achieve different levels of energy/performance trade-off depending on the application. The next task that is scheduled depends on the experiences of the node

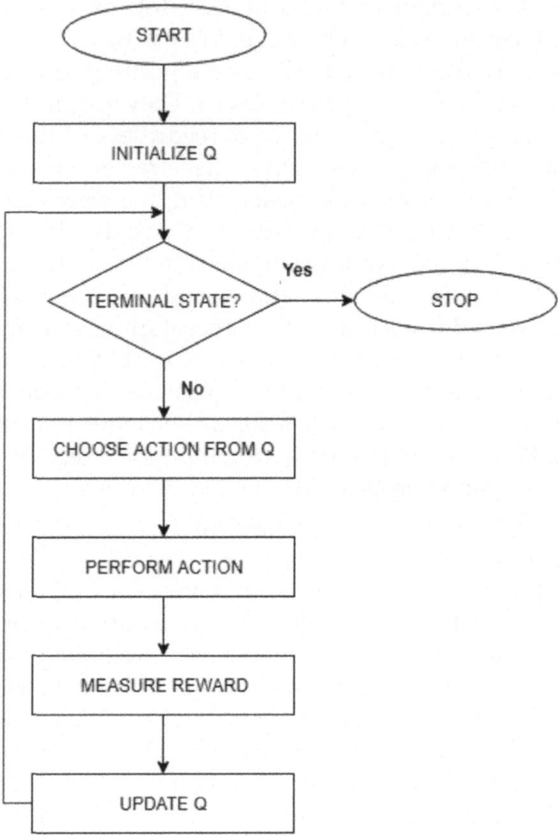

FIGURE 3.7
Q learning flowchart.

so far. The algorithm is applied to the case of target tracking in WSNs and cooperation between neighboring nodes, which combined with an RL approach is used to significantly reduce the energy consumption for tracking targets in the network. Experimentation was carried out by the authors in a C# simulation environment and promising results were obtained. It was noted that their cooperative RL approach outperformed non-cooperative approaches in terms of tracking quality. Q-learning has also been successfully deployed for the case of routing in WSNs, which has been explained in Section 3.2.2.

3.2.1 Artificial Intelligence in WSN

There have also been many other applications of artificial intelligence techniques in WSNs. Some of the popular applications have been mentioned in short here too. In any WSN, the network lifetime is of primary consideration, as each sensor has limited computation, energy and memory resources. Appropriate cluster-head election could significantly reduce the energy consumption, as well as enhance the lifetime of a network. In [31], the authors have proposed the use of fuzzy logic to make this election. The three fuzzy descriptors that were used in the fuzzy rule base are node concentration, the energy reserve of each node and centrality of the node with respect to the entire cluster. The output of the fuzzy model is the chance that node has of becoming the cluster-head. The proposed model was tested on three sample networks with medium-sized clusters. It was found that using this system model, a significant increase in network lifetime was achieved. Also, further accurate modifications of each fuzzy sets could yield better results.

In [32], the researchers have proposed the use of the soft computing technique fuzzy logic to enhance the lifetime of the network. Since individual sensors have limited battery lives and replacing or recharging them isn't cost effective. Thus, WSNs are primarily used as 'use and throw networks' to monitor the environment in which they are dispersed. In such a scenario the efficient use of energy resources is crucial. The authors have proposed the use of fuzzy logic to divide the network into sections called clusters, with each cluster electing a leader from among its members. Only these leaders communicate with the base station. To select the cluster-heads, two descriptors are used as the input for the fuzzy logic system: required energy and residual energy. The output of the fuzzy rules is the possibility that a node becomes a cluster-head. The proposed clustering technique is divided into two steps. In the first step, the network base station broadcasts 'Hi' message to the network. Then individual sensor nodes calculate the required energy to transmit k bits to the base station according to the signal strength received. Thus, the nodes now have the values for the input to initialize the fuzzy logic system (with residual energy and required energy). The second step is similar to the LEACH clustering technique. The proposed methodology is compared with LEACH and CHEF clustering approaches and found to outperform them in terms of energy conservation. Also, this process can be implemented in both homogeneous as well as heterogeneous WSNs.

As is shown in previous research, WSNs employ clustering to increase the overall lifetime of the network. Power consumed by a sensor node to transmit data is the most important factor that affects network lifetime severely. Thus, decreasing the long-distance communication between sensors and sink nodes will lead to a better network performance. Clustering is the best method for reducing this. In [33], the researchers have proposed a model to incorporate the SC technique genetic algorithm (GA) with hierarchical clustering to reduce these long-distance communications, as given in Figure 3.8. In GA, the chromosome is represented as a list of parameters called genes. In the case of WSNs, the chromosome is the network topology. In the proposed methodology, an initial population of chromosomes is initialized and their fitness is found based on preset parameters, which are the network lifetime. Operations like selection, crossover and mutation are performed iteratively

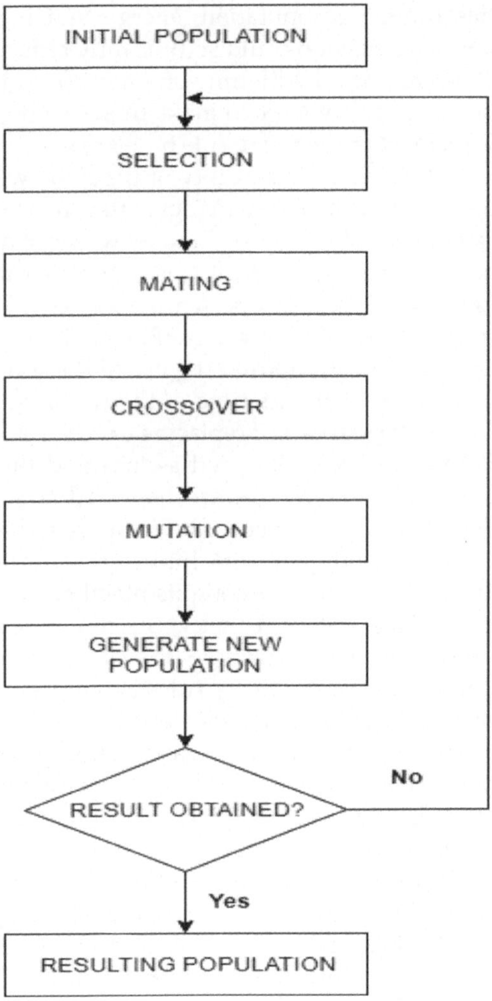

FIGURE 3.8
Genetic algorithmic approach.

to increase the fitness of the next generation. As soon as the fitness of a solution reaches near-optimum levels, the process is stopped and the computed nodes are assigned as cluster-heads. The authors simulated the performance of GA in the case of the clustering problem for WSNs and found that the solution was reached efficiently without any complex computations. Thus, the proposed method is said to be an efficient method to find cluster-heads. Also, the approach significantly reduced long-distance exchanges as well. It is also posited to be a scalable approach and can be implemented for a large number of nodes with varying base station locations and deployment styles of nodes.

Not only should the WSNs nodes have efficient task scheduling at each sensor node, but they should also be able to work cooperatively to solve complex problems together. In [34], the authors have proposed the use of the SC technique of swarm optimization to solve the problem of task allocation. They have proposed a modified version of binary particle swarm optimization (MBPSO) that uses a new transfer function and position updating procedure (with mutation) to get the optimum solution. Parameters like the total energy consumption, total task execution time and energy distribution are used as the cost functions for the task allocation problem. Each particle's velocity, as well as binary position, is updated depending on the fitness to generate a new swarm. Mutation is also carried out to randomly insert new particles that might not be otherwise generated. Simulations were carried out by the researchers and their proposed MBPSO–based approach was found to be feasible. It was also found to outperform approaches solely based on GA and BPSO paradigms. However it is not always the case that SC techniques would yield better results. Especially in the case of maximizing the area coverage of WSNs, algorithms like improved cuckoo search and chaotic flower pollination (as explained in [35]) have been shown to outperform democratic particle swarm optimization (DPSO) and particle swarm optimization (PSO) by a huge margin.

3.2.2 Routing Using the Machine Learning Approach in WSNs

The use of ML techniques in WSNs has gained importance due to the huge versatility they offer. A key area in which ML techniques are needed the most is in the routing protocols used by the sensor nodes; both amongst themselves and in the base stations. WSNs vary a lot compared to traditional ad hoc networks in terms of the higher quantity of sensor nodes, higher failure rates than ad hoc, non-rechargeable battery capabilities, limited computational capabilities, etc. Due to these constraints, traditional routing protocols would prove inefficient when applied for the problem of packet transmissions in WSNs. Two of these techniques are flooding, in which the node broadcasts the incoming packet through every outgoing link (except the one where it was sent from) and gossiping, where the node would randomly select a node to send the incoming packet to. These steps are carried on until all the nodes receive the packet. Both these techniques would result in several drawbacks when applied to WSNs. Flooding would not only waste the limited battery capabilities of each sensor node, but would also result in implosion and overlap. Whereas, gossiping would result in a high delay for packet reaching the destination. Thus, WSNs require a customized routing policy to ensure they perform efficiently when the sensor nodes are deployed.

In [36], the authors have explored the use of soft-computing techniques to design routing models for WSNs. these models focus on optimizing the network lifetime (as each node has a limited battery life). Soft computing techniques like fuzzy logic (FL), neural networks (NN), swarm intelligence, genetic algorithms, and reinforcement learning have been studied and examined by them to optimize routing. The parameters that were given primary consideration were power efficiency, network lifetime, path discovery, routing optimization, and computation requirements. The authors have compared the mentioned SC techniques and formulated their findings in a tabular format. It is noteworthy to find that NN was found to be the best technique in terms of figuring out the best path (path discovery), however, it fared badly when it came to computation required.

In [37], the authors have attempted to develop an energy efficient routing protocol for WSNs based on fuzzy logic. The energy-aware routing scheme is based on a cross-layer approach to minimize the overall energy consumed and thus maximize the network lifetime. The inputs to the fuzzy logic system are the remaining battery, received signal strength and transmission power for nodes within local communication range. These parameters determine the next hop relay node that the sensor must send the data to for reaching the sink node. A disadvantage of a non-dynamic system is that if a sensor node probabilistically keeps on selecting the same node to relay data to the sink node and ignores other alternatives, then the power level of that node depletes much quickly than the others. The proposed model dynamically selects different nodes according to the fuzzy system decision and thus extends the lifetime of the entire WSNs as a whole. The authors ran simulation based on their proposed model and their results yielded positive results with respect to extended network lifetime capabilities.

One of the main constraints to be considered when developing a routing policy for WSNs is that the policy should attempt to keep the communication and computational overhead at the minimum while still performing at near-optimum levels. It was concluded in [38] that Q-learning algorithms were able to achieve that. However, the original Q-learning approach failed to perform efficiently when it switched from a high traffic load to a low traffic load. Instead of adapting itself to use the shortest path policy, it kept on using the load balancing approach.

This has been rectified by the researchers in [39] by modifying the Q-learning approach for WSNs. They have proposed an algorithm that balances the energy expenditure in WSNs by taking into consideration the neighbors' energy levels when updating the nodes' routing table. It does do this by generating a multiplier that depends on this energy level. When updating that neighboring nodes' Q-value in the routing table, a weighting function is used. The lower the energy level, the higher the multiplier and thus further communication via that route is discouraged. The authors ran simulations based on their proposed algorithm and it was found that their energy-aware, metrics-based model would always increase the lifetime of the network. This was the case for both homogeneous- and heterogeneous-composed topologies.

To tackle the limitations of WSNs, researchers in [40] have proposed the use of ultra-wide band (UWB) [41] technology for developing a reinforcement learning based geographic routing protocol (RLGR) for UWB sensor networks. The authors assumed that the nodes are equipped with a UWB transceiver and are stationary, know their location, and know their neighbor's location and residual energy. Using these preconditions, the authors have proposed the RLGR model to increase the network lifetime. They have introduced a novel reward function for the Q-learning-based approach in routing. The model takes into consideration four items: a normalized advance to the sink in next hop and energy required for that hop; reward for reaching sink in one hop; solving the 'void' problem in geographic routing by using negative reward mechanism and using low energy warning signals. After sending out the packet, the node calculates its reward for its selection using this reward function. It then updates the Q values for its neighbors using the calculated reward. The authors ran simulations and compared their findings to the popular GPSR [42] routing technique. It was found that RLGR outperformed GPSR in terms of network lifetime for both grid and randomly distributed networks. However, it had a higher delay for grid networks and lower delivery ratio for randomly distributed networks.

3.3 Conclusion

Wireless sensor networks entail the use of various sensors scattered over a region and are thus different from traditional networks in various aspects. They necessitate the use of techniques that are unique for their individual applications. Wireless sensors deployed in underwater regions have vastly different protocols than those deployed in forest regions. Thus, the various problems facing WSNs in different conditions have become a hot topic for researchers. The emergence of Internet of Things has further increased interest into developing systems that extend the capabilities of sensor nodes. Promising work is being done, not just in terms of using WSNs to monitor the surroundings of human beings, but also in using sensors to monitor the internal health of human beings as shown in [43].

Chapter 4 highlights how machine learning algorithms have been adopted for increasing the capabilities of WSNs. The impact of these algorithms to enhance the abilities of WSNs to adapt to dynamic and ever-changing environmental conditions have been discussed. In this age of ubiquitous data, supervised learning techniques have been given a lot of attention by researchers. Moreover, with the resurgent popularity of neural networks and easily available fast computing power, a lot of new avenues have opened up for using the supervised learning paradigms to WSNs. However, other techniques like clustering (unsupervised learning) and Q-learning (reinforcement learning) have also firmly established themselves in solving niche problems in WSNs.

Although a lot of promising work has already been done, numerous issues in WSNs are still unsolved and require further dedicated research. Especially in solving problems like developing lightweight-message passing protocols, hierarchical clustering, distributed processing and using machine learning for resource management of wireless sensor networks.

References

1. Nilanjan Dey, Amira S. Ashour, Fuqian Shi, Simon James Fong, and R. Simon Sherratt. "Developing Residential Wireless Sensor Networks for ECG Healthcare Monitoring." *IEEE Transactions on Consumer Electronics* Vol. 63, No. 4, November 2017.
2. Tao-Yang Fu, Wen-Chih Peng, and Wang-Chien Lee. "Parallelizing Itinerary-Based KNN Query Processing in Wireless Sensor Networks." *IEEE Transactions on Knowledge and Data Engineering* Vol. 22, No. 5, May 2010.
3. H. Schwetman, CSIM User's Guide (Version 18), Mesquite Software, Inc., http://www.mesquite.com, 2009.
4. S.H. Wu, K.T. Chuang, C.M. Chen, and M.S. Chen. "DIKNN: An Itinerary-Based KNN Query Processing Algorithm for Mobile Sensor Networks." *Proc. 23rd IEEE Int'l Conf. Data Eng. (ICDE)* 456–457, 2007.
5. J. Winter, Y. Xu, and W.-C. Lee. "Energy efficient processing of k nearest neighbor queries in location-aware sensor networks." 2nd International Conference on Mobile and Ubiquitous Systems: Networking and Services. IEEE, 2005, 281–292.
6. P. P. Jayaraman, A. Zaslavsky, and J. Delsing. "Intelligent processing of k-nearest neighbors queries using mobile data collectors in a location aware 3D wireless sensor network." *Trends in Applied Intelligent Systems.* Springer, 2010, 260–270.
7. Aymen Abid, Awatef Ben Fradj Guiloufi, Nejah Nasri, Abdennaceur Kachouri, Adel Mahfoudhi, and Mohamed Abid. "Centralized KNN anomaly detector for WSN." 2015 12th International Multi-Conference on Systems, Signals & Devices.
8. G. Ahmed, N.M. Khan, Z. Khalid, and R. Ramer. "Cluster head selection using decision trees for wireless sensor networks." International Conference on Intelligent Sensors, Sensor Networks and Information Processing. IEEE, 2008, 173–178.
9. Vinicius F. Silva, Daniel F. Macedo, Jesse L. Leoni. "Machine Learning-Based Spectrum Decision Algorithms for Wireless Sensor Networks." 2016 13th IEEE Annual Consumer Communications & Networking Conference (CCNC).
10. Ashish Payal, C.S. Rai, and B.V.R. Reddy. "Artificial Neural Networks for Developing Localization Framework in Wireless Sensor Networks." International Conference on Data Mining and Intelligent Computing 2014 (ICDMIC).

11. A. Shareef, Y. Zhu, and M. Musavi. "Localization using neural networks in wireless sensor networks." *Proceedings of the 1st International Conference on Mobile Wireless Middleware, Operating Systems, and Applications* 2008, 1–7.
12. S. Yun, J. Lee, W. Chung, E. Kim, and S. Kim. "A soft computing approach to localization in wireless sensor networks." *Expert Systems with Applications* Vol. 36, No. 4, 7552–7561, 2009.
13. S. Chagas, J. Martins, and L. de Oliveira. "An approach to localization scheme of wireless sensor networks based on artificial neural networks and genetic algorithms." 10th International Conference on New Circuits and Systems. IEEE, 2012, 137–140.
14. Zhilbert Tafa. "Artificial Neural Networks in WSNs Design: Mobility Prediction for Barrier Coverage." 2016 IEEE International Symposium on Signal Processing and Information Technology (ISSPIT).
15. H. He, Z. Zhu, and E. Makinen. "A neural network model to minimize the connected dominating set for self-configuration of wireless sensor networks." *IEEE Transactions on Neural Networks* Vol. 20, No. 6, 973–982, 2009.
16. L. Yu, N. Wang, and X. Meng. "Real-time forest fire detection with wireless sensor networks." *International Conference on Wireless Communications, Networking and Mobile Computing* Vol. 2, 2005, 1214–1217.
17. R.V. Kulkarni and G.K. Venayagamoorthy. "Neural network based secure media access control protocol for wireless sensor networks." *Proceedings of the 2009 International Joint Conference on Neural Networks* ser. IJCNN'09. Piscataway, NJ, USA: IEEE Press, 2009, 3437–3444.
18. A. Snow, P. Rastogi, and G. Weckman. "Assessing dependability of wireless networks using neural networks." *Military Communications Conference* IEEE, 2005, 2809–2815 Vol. 5.
19. Woojin Kim, Jaemann Park, Jaehyun Yoo, H. Jin Kim, and Chan Gook Park. "Target Localization Using Ensemble Support Vector Regression in Wireless Sensor Networks." IEEE Transactions on Cybernetics.
20. Tran and T. Nguyen. "Localization in wireless sensor networks based on support vector machines," *IEEE Transactions on Parallel and Distributed Systems.* Vol. 19, No. 7, 981–994, 2008.
21. B. Yang, J. Yang, J. Xu, and D. Yang. "Area localization algorithm for mobile nodes in wireless sensor networks based on support vector machines." *Mobile Ad-Hoc and Sensor Networks* Springer, 2007, 561–571.
22. Carlos Giovanni Nunes de Carvalho, Danielo Gonçalves Gomes, José Neuman de Souza, and Nazim Agoulmine. "Multiple Linear Regression to Improve Prediction Accuracy in WSN Data Reduction." Network Operations and Management Symposium—2011 7th Latin American (LANMOS).
23. Christiana Ioannou, Vasos Vassiliou, and Charalampos Sergiou. "An Intrusion Detection System for Wireless Sensor Networks." 24th International Conference on Telecommunications—2017 (ICT).
24. Tianyu Zhang, Qian Zhao, and Yukikazu Nakamoto. "Faulty Sensor Data Detection in Wireless Sensor Networks Using Logistical Regression." 2017 IEEE 37th International Conference on Distributed Computing Systems Workshops.
25. Pedro A. Forero, Alfonso Cano, and Georgios B. Giannakis. "Distributed Clustering Using Wireless Sensor Networks." *IEEE Journal of Selected Topics in Signal Processing* Vol. 5, No. 4, August 2011.

26. A. Rooshenas, H. Rabiee, A. Movaghar, and M. Naderi. "Reducing the data transmission in wireless sensor networks using the principal component analysis." 6th International Conference on Intelligent Sensors, Sensor Networks, and Information Processing. IEEE, 2010, 133–138.
27. S. Macua, P. Belanovic, and S. Zazo. "Consensus-based distributed principal component analysis in wireless sensor networks." 11th International Workshop on Signal Processing Advances in Wireless Communications, 2010, 1–5.
28. C. Fenxiong, L. Mingming, W. Dianhong, and T. Bo. "Data compression through principal component analysis over wireless sensor networks." *Journal of Computational Information Systems* Vol. 9, No. 5, 1809–1816, 2013.
29. C. Watkins and P. Dayan. "Q-learning." *Machine Learning* Vol. 8, No. 3–4, 279–292, 1992.
30. Muhidul Islam Khan and Bernhard Rinner. "Energy-aware Task Scheduling in Wireless Sensor Networks based on Cooperative Reinforcement Learning." Workshop on Energy Efficiency in Wireless Networks & Wireless Networks for Energy Efficiency (ICC'14).
31. Indranil Gupta, Denis Riordan, and Srinivas Sampalli. "Cluster-head Election using Fuzzy Logic for Wireless Sensor Networks." 3rd Annual Communication Networks and Services Research Conference (CNSR'05).
32. Manjeet Singh, Surender Soni, Gaurav, and Vicky Kumar. "Clustering using Fuzzy Logic in Wireless Sensor Networks." 3rd International Conference on Computing for Sustainable Global Development 2016 (INDIACom).
33. Mohammad M. Shurman, Mamoun F. Al-Mistarihi, Amr N. Mohammad, Khalid A. Darabkh, and Ahmad A. Ababnah. "Hierarchical Clustering Using Genetic Algorithm in Wireless Sensor Networks." 36th International Convention on Information and Communication Technology Electronics and Microelectronics 2013 (MIPRO).
34. Jun Yang, Hesheng Zhang, Yun Ling, Cheng Pan, and Wei Sun. "Task Allocation for Wireless Sensor Network Using Modified Binary Particle Swarm Optimization." *IEEE Sensors Journal* Vol. 14, No. 3, March 2014.
35. Huynh Thi Thanh Binh, Nguyen Thi Hanh, La Van Quan, and Nilanjan Dey. "Improved Cuckoo Search and Chaotic Flower Pollination optimization algorithm for maximizing area coverage in Wireless Sensor Networks." *Neural Computing and Applications* 1–13.
36. Vijayalaxmi Kadrolli and Jayashree Agarkhed. "Soft Computing Routing Techniques in Wireless Sensor Network." International Conference on Advances in Electrical, Electronics, Information, Communication and Bio-Informatics (AEEICB16).
37. Toleen Jaradat, Driss Benhaddou, Manikanden Balakrishnan, and Ala Al-Fuqaha. "Energy efficient cross-layer routing protocol in Wireless Sensor Networks based on fuzzy logic." 9th International Wireless Communications and Mobile Computing Conference 2013 (IWCMC).
38. A. Forster. "Machine learning techniques applied to wireless adhoc networks: Guide and survey." Proceedings of the 3rd International Conference on Intelligent Sensors, Sensor Networks and Information Processing (ISSNIP), ISSNIP, 2007.

39. Kaveri Kadam and Navin Srivastava. "Application of Machine Learning (Reinforcement Learning) for Routing in Wireless Sensor Networks (WSNs)." 1st International Symposium on Physics and Technology of Sensors 2012 (ISPTS).
40. S. Dong, P. Agrawal, and K. Sivalingam. "Reinforcement learning based geographic routing protocol for UWB wireless sensor network." Global Telecommunications Conference. IEEE, 2007, 652–656.
41. X. Sheng, M. Guizani, R.C. Qiu, and T. Le-Ngoc. *Ultra-Wideband Wireless Communications and Networks*. Wiley, June 2006.
42. B. Karp and H.T. Kung. "GPSR: Greedy perimeter stateless routing for wireless networks." *Proc. ACM/IEEE MobiCom* 243–254, Boston, Mass., USA. August 2000.
43. A.S. Ashour, F. Shi, and R.S. Sherratt. IEEE Reviews in Biomedical Engineering Methodological Reviews Title "Wireless Capsule Gastrointestinal Endoscopy: Direction of Arrival Estimation based Localization Survey."

4

Relevance of Soft Computing Techniques in the Significant Management of Wireless Sensor Networks

Mamata Rath, Bibudhendu Pati, and Binod Kumar Pattanayak

CONTENTS

4.1 Introduction

In recent technological development, soft computing approaches play an important role to solve various problems based on approximation and probability. If we strategize to solve such issues, gradually the hard computing paradigms are supplanted by soft computing methods, such as artificial neural networks (ANN) [1], genetic algorithms (GA) [2], fuzzy logic systems [3], and particle swarm intelligence. Data mining and data modeling techniques are mostly used for refining data when data aggregation methods do not follow proper data models and there are many operational errors. The precision of the information can be expanded by utilizing the data combination and the accumulation operations. Accompanied by the numerous repetition of similar information, or when information is produced from different sources, precision must be expanded [4]. Parameters, for example, way length and vitality utilization, are required for outlining multi-way calculation. A combinational streamlining issue is taken for the enhancement of the network parameters for WSNs directing procedure. This issue can be illuminated by utilizing the soft computing strategy of ant colony optimization (ACO) [5].

Circulated computing, self-association, and positive criticism incorporates the qualities of the ACO calculation. Current correspondence networks utilize ACO for course seeking [6]. The nature of data can be enhanced, by utilizing fluffy rationale techniques which can combine questionable information from numerous sensor hubs. The ordinary scientific computational techniques (for example, expansion, subtraction, and addition) and division require more computational power when contrasted with fluffy rationale strategies. The final precise outcomes can be found out by utilizing just couple of tests. The issues are portrayed by human experts and are more efficiently controlled [7]. Using genetic algorithm techniques, energy efficient information aggregation spanning trees are created in wireless sensor networks. These trees are identified by chromosomes and the value of gene significance and node number specifies the parent node identified by gene index. The customary genetic activities such as single-point crossover, mutation, tournament choice, and eclecticism are used for the following production purposes [8]. At the point when an introverted sensor node utilizes all its energy and the remaining sensors expend the remaining energy, the WSN's lifetime gradually reduces. Efficient approaches to dispense undertakings to the hubs can be accomplished by anticipating the significant hubs and this expansion the lifetime of the network [9].

4.2 An Overview of Wireless Sensor Networks (WSNs)

Wireless sensor networks (WSNs) are comprised of spatially disseminated, self-governing sensors fit for restricted computing power, storage room, and a correspondence range to screen physical or ecological conditions (for example, temperature, sound, and weight) and to agreeably pass on their information through the network to a fundamental area. The improvement of WSNs was roused by military applications (for example, front line reconnaissance). Today, such networks are utilized as a part of numerous mechanical applications, such as modern and assembling computerization, assessing the well-being of machines, etc. When utilizing a WSN as an exploratory foundation, it is frequently necessary to gather covered structures in distributed information gathered by the sensor nodes. For a scene division and observing application, nodes in the WSN need to team up with each other so as to fragment the scene, distinguish the protest of intrigue, and characterize it from that point [10]. Anomaly location is likewise an ordinary assignment for WSNs with applications [11,12] in checking synthetic spillage and interruption recognition. Nodes in a WSN group bind together to form clusters and perform a similar type of operation. The same category of applications can be applied. Numerous implemented clustering techniques have been proposed by network research groups to tackle the information grouping issue, keeping in mind the end goal is to get a handle on the crucial parts of the network life time.

In any case, a lot of information is dispersedly gathered in topographically conveyed nodes over networks in WSNs [13]. Because of the restricted vitality, correspondence, methodology, and capacity assets, incorporating the entire circulated information [14,15] to one combination node to perform brought together grouping might be unrealistic. In this manner, there is extraordinary interest in conveyed information grouping methodologies, in which the worldwide bunching issue can be fathomed at every individual node in light of nearby information and restricted data trades among nodes. Moreover, contrasted and the incorporated bunching, circulated grouping is more adaptable and strong to nodes and additionally connect disappointments. Clustering in WSN [16] can be accomplished by different methodologies that contrast essentially in their thought of what constitutes a group and how to productively discover a group. A standout amongst the most mainstream ideas of the bunches is bunches with little in-gathering [17,18] and substantial out-assemble contrasts. Diverse measures of distinction (or likeness) might be utilized to put things into groups [19] including separation and force. A lot of methodologies have been proposed to manage the grouping issue in view of separation. Among them, the k-means methodology is the most well-known hard bunching methodology, which highlights basic and quick joined cycles.

WSNs are an assembly of a large number of self-sorted out kinds of sensor nodes which collaborate together to monitor and record network routing, communication, and transmission. (i.e., they areused to quantify temperature, sound, and weight) and pass accumulated data to the central area. WSNs nodes and group of stations communicate between real and virtual conditions, which makes them more useful for some applications [10]. Chiefly, WSNs are utilized in the military, yet nowadays they are utilized in different areas like mechanical applications, purchaser applications, human services, among others. In spite of having numerous focal points, there are a few issues that are occuring in WSNs like hotspot issues, vitality opening issues, steering, scope issues, load balancing issues, etc. These issues impact various components [17,18] of WSNs, namely vitality utilization, dependability, quality, organization time, and the lifetime of networks, which corrupt the execution of WSNs. To unravel these issues different scientists have created distinctive components. Among them [6], a diverse sort of soft computing standards has been studied. There are different sorts of soft computing methods created: swarm knowledge, fluffy rationale, neural networks, fortification learning, and transformative methodology, all of which are used to tackle WSN issues with the goal that execution of the network will be expanded. Information aggregation is another challenging issue in WSNs and is currently being carried out using various data mining approaches.

In WSNs, information aggregation with soft computing strategies is a demanding issue from a security point of view. When a node is compromised, it is easy for an opponent intruder node to spread false information and manipulate the aggregator into acknowledging false readings. Along these lines, there is a requirement for secure information. An itemized study [8] presents real commitments to the security arrangements in information conglomeration which predominantly utilizes soft computing procedures. In [8], a classification of conventions are finished by the soft computing method, for example, fluffy rationale, swarm knowledge, hereditary methodology, and neural networks [20]. Precision, vitality utilization [21,22], cost decrease, and safety efforts are the measurements utilized for classification.

4.3 An Overview of Soft Computing Applications and Their Relevance in WSNs

Section 4.3 describes in brief the various application of soft computing technique in WSNs. One such application area is traffic light management in intelligent transport systems, where the challenge lies in correct management of traffic light signals due to the collected dynamic flow of various vehicles. The authors in [23] presented an innovative approach called neverstop, using a genetic algorithm and fuzzy logic process in a sharp traffic organization system. Neverstop has been developed using sensors to operate the traffic lights at intersections. To adjust the timing intervals of traffic lights, fuzzy control mechanisms and genetic algorithms have been used in a way that reduces the waiting time for the vehicles at the traffic post. A model called EBox-II uses a peripheral device that executes the fuzzy control system. It has been experimentally proven that neverstop performs efficiently with gradual waiting time reduction for traffic flow. There has always been a challenge for search engines to extract information from the Internet dynamically and intelligently in order to reduce search time and to present an optimized result for end users. The authors in [24] presented a better transport design and its model named as ITTS 9 (Intelligent Transport and Tourism System) using Hadoop.

The developing technology in WSNs has various applications in different fields, for example, healthcare, the military, business, robotics, and so on. This is conceivable because of a couple of unmistakable characterizations of WSNs. For example, they have the ability of detecting different flags at the same time in a financially savvy way, yet a few requirements like the dependable steering of hubs and localization without GPS (global positioning system). A GPS can be effectively utilized in open territories. With the "got flag" feature, the evaluated separation can be inferred in a localization technique that does not rely upon scope of the flag. This requires an enhancement approach in light of soft computing procedures for more power and an easier interface. Soft computing techniques depend on approximation, vulnerability, and fractional truth. The authors of [9] demonstrated the material soft computing systems in localization issues and show another enhanced approach by using an outrageous learning machine to upgrade the rate of exact approximation. To enhance the accuracy more, an insatiable approach coordinated the adjusted mass spring advancement with a covetous technique, which showed the viability of the blended approach when contrasted and other comparable methodologies.

4.3.1 Intelligent Application of Soft Computing in WSNs

Cloud computing is emerging rapidly as a client–server technology with various service applications. Nonetheless, given the accessibility of a various scope of remote access innovations, individuals anticipate that persistent association will become the most appropriate innovation. The authors in [25] presented two soft computing localization techniques for wireless sensor networks called neural fuzzy inference system (ANFIS) and ANN, which focus on a range-based localization technique that depends on the measurement of the arriving signal strength indicator (RSSI) from the three ZigBee anchor nodes disseminated throughout the track cycling field. The soft computing techniques aimed to estimate the distance between bicycles touching on the cycle track for outdoor and indoor velodromes. In the first approach, the ANFIS was considered, whereas in the second approach, the ANN was hybridized individually with three optimization algorithms, namely particle swarm optimization (PSO), gravitational search algorithm (GSA), and backtracking search algorithm (BSA). The results revealed that the hybrid GSA-ANN outperformed the other methods adopted in this paper in terms of accuracy, localization, and distance estimation accuracy.

Classification schemes in artificial neural networks are discussed in [25] with three optimization algorithms, namely PSO, GSA, and BSA.

Among the fundamental difficulties of present day communication is the availability of remote systems that use cell phones. Previous research has shown a few heuristic methodologies that rearranged standards and indicated the best system possible. Cell phones require calculations that are easy to use when considering the best accessible system close constant. A comprehensive study in [5] proposed a hybrid intelligent handover decision algorithm principally founded on two main heuristic algorithms: ABC and PSO, named ABC-PSO in order to choose the best remote system amid the vertical handover process. The ABC-PSO calculation has been streamlined to accomplish low-cost work that comply with IEEE 802.21 standards, considering diverse accessible remote systems, the application necessities, and the

TABLE 4.1

Intelligent and Smart Applications in WSN Using Soft Computing and Related Studies

Sl.No	Literature	Year	Application area of Soft Computing in WSN
1	Chao Wang et al. [23]	2016	Big data intelligent transport system
2	T. Raghunathan et al. [24]	2015	Intellectual transport and traffic system
3	Chakchai et al. [9]	2016	Localization trouble solution in WSNs
4	Le Hoang et al. [26]	2017	Wi-Max Network Planning of 3D GIS
5	Goudarzi et al. [5]	2017	Ant bee colony, PSO in handover practice of WSNs
6	S. Gheisari et al. [27]	2017	Cognitive WSN based on Bayesian network
7	G. Pachauri et al. [28]	2015	Anomaly finding in WSN using machine learning
8	S. Rashid et al. [29]	2015	WSN size calculation using machine learning
9	S. Gheisari et al. [30]	2016	Learning automata-based cognitive WSN
10	J. Wang et al. [31]	2017	PSO-based clustering algorithm in WSN
11	P.P. Patel et al. [6]	2016	Survey on soft computing tools applied in WSN
12	S.J. Habib et al. [32]	2014	Prevention and restoration of failure in WSN

client inclinations to enhance Quality of Service (QoS). Numerical outcomes show that the ABC-PSO calculation contrasted with the related work has bring down cost and deferral, has shown a higher accessible data transmission. Table 4.1 shows intelligent applications in WSNs using soft computing and related studies.

The use of delicate computing techniques for the issue of WiMax (Worldwide Interoperability for Microwave Access) Network Planning [26] on 3D geographical information systems (3D GIS) that upgrades both execution of the system (coverage and QoS) and venture costs (the quantity of base stations and parts) which have been developed in [26]. A preparation technique using the most recent aftereffects of parallel random forest order calculation to decide substantial places of construct stations in light of a landscape of 3D GIS was proposed. To produce an optimal resolution of the model, a hybrid algorithm between greedy BTP and enhanced particle swarm optimization included with parallel computing is presented. Investigational validation of the planned method in assessment with other significant ones is performed.

4.3.2 Neural Network Modeling in WSNs

ANFIS was considered in an innovative approach [25], whereas in the second approach the ANN was hybridized individually with three optimization algorithms, namely PSO, GSA, and BSA. The plan of present day wireless network-based systems, which include basic intelligent decision building and parameter optimization, is very difficult because of the exceptionally powerful, and regularly obscure, natural conditions that describe remote systems. There is a typical pattern in current systems to consolidate artificial intelligence (AI) methods to adapt to this outline unpredictability. While various AI systems have been gainfully utilized in the remote systems group, the entrenched AI framework of neural networks (NNs), understood for their striking all inclusive statement and adaptability, has been connected in a wide assortment of settings in remote systems. Specifically, NNs are particularly prominent for errands including characterization, learning, or optimization.

The authors in [1] give both an article of regular NN models and a far-reaching study of the utilizations of NNs in remote systems. It distinguishes the entanglements and difficulties of actualizing NNs. Recent wireless communication technologies empower the organization of extensive scale sensor networks. Sensor nodes have restricted power energy requirements which is the basic issue in the organization. Subsequently, we should create energy efficient techniques to encourage effective information transmission between the nodes. In [4], there is an examination of present accessible energy proficient clustering strategies. Likewise, it proposes a dynamic clustering node strategy utilizing a neural network. In NNs every node needs to send information to each node in the following layer, which in turns requires heaps of energy. In each cluster the network chooses the cluster node. Cluster nodes form the information and send it to the base station or through other cluster nodes. Intermittently, the cluster and cluster node are changed, keeping in mind the end goal to adjust the energy limitations.

4.3.3 Fuzzy Logic-Based Intelligent Solution Strategy in WSNs

Mine fire is one of the real perils in underground coal mines. The mine's security is consistently dependent on surveying the mines for flames. Currently, WSNs are used in this regard; the sensor nodes that are conveyed in coal mines for gathering different kinds of ecological information, for example, temperature, relative mugginess, centralizations of various gases, and so forth. This information is then sent to a base station (sink) straightforwardly associated with the ground checking center. Since the checking information is questionable and uncertain in nature, it is vital to decipher and investigate the information before taking measures for fire prevention. For the authors in [3], a WSN-based fire observing framework for underground coal mines has been displayed utilizing a fluffy rationale approach for improving the unwavering quality in basic leadership procedure to ease the mine fire risk. The Mamdani fluffy deduction framework is utilized as fluffy model for verifying information.

The framework is widely reproduced using the fuzzy logic toolbox in MATLAB®. The recreation comes about uncovered that the proposed framework is more dependable and receptive to flame peril when contrasted with the disconnected observing framework utilized as a part of underground coal mines.

One of the key challenges is correct monitoring of the position of the node. It permits successful information transmission between sensor nodes, keeping in mind the end goal to spare their vitality and broaden the system lifetime. The authors in [33] proposed and actualized another instrument for geographic directing. Accordingly, the proposed system was dependent on a weighted centroid limitation procedure, where the places of obscure nodes were ascertained utilizing the fluffy rationale technique. For this, a fluffy restriction calculation was utilized and received a stream estimation through a remote channel to register the separation between the stay and the sensor nodes. Along these lines, our work depends on the centroid calculation that computes the position of obscure nodes utilizing fluffy Mamdani and Sugeno surmising frameworks for expanding the precision of assessed positions. Once the restriction calculation has identified the area of nodes with obscure position, the proposed component chooses the following cluster head to lessen the vitality scattering of sensor nodes, which prompts an expansion of the system lifetime. The three primary favorable circumstances of the proposed system are to limit the position blunder of nodes and diminish the mistake restriction normal; to expand the quantity of parcels transmitted to the following bounce cluster head in view of the restriction calculation; and to diminish the vitality utilization of nodes and after that expands the system lifetime utilizing an effective determination of next group cluster head.

One of the most recent ideal models around there is programming characterized by the wireless sensor network. As indicated by the researchers, the systems deal with overseeing topological data and sending choices utilizing a bipartite design in which a control plane chooses the sending approaches and the information plane (i.e., common sensor nodes) and executes them. Sadly, in exceptionally powerful systems, this approach creates an overhead of control parcel trade between the standard nodes and the control plane, that prompts extra vitality utilizations. The authors in [7] proposed a fluffy rationale-based arrangement, called fuzzy topology discovery protocol (FTDP), to enhance the effectiveness of programming characterized by remote sensor systems. The work is composed by the software defined networking answer for wireless sensor systems (SDN-WISE), which is an open source answer for programming characterized by remote sensor systems. The proposed work is one of the primary endeavors to utilize the fluffy hypothesis in programming characterized by remote sensor systems. The reproduction demonstrates that this approach can build the lifetime of the system by 45% and diminishes the parcel misfortune proportion by half contrasted with the fundamental SDN-WISE arrangement.

In the past few years, several methodologies were executed with clustering of the nodes in wireless sensors networks. Each of these calculations has its own particular advantages and challenges. The common denominator of all these methodologies is the steadiness of the calculation in every one of the rounds of the system lifetime that causes the determination of group heads in each round [34]. By contrasting the possibility of every node that will be chosen as a group head utilizing an irregular number, the dominant part of these bunching approaches, both fluffy and non-fluffy, diminish the chances of choosing the most qualified node as the bunch head. Therefore, all these methodologies require the choice of group heads in each round. Choosing group heads in each round builds the measure of received and sent messages to such an extent that in systems with substantial number of nodes, it causes a few issues, for example, vitality diminishment, improvement of impact, and system movement. Be that as it may, by choosing the most qualified nodes as group heads for no less than a couple of rounds, the measure of sent and received messages decreases. In [35], a versatile multi-clustering calculation utilizing fluffy rationale in a remote sensor organizer (adaptive MCFL) is displayed. Notwithstanding, grouping nodes in various rounds utilizing distinctive bunching calculations, the proposed calculation abstains from choosing new group heads by trusting past groups. The proposed approach is contrasted and different methodologies in three distinct situations utilizing records, for example, remaining vitality and the quantity of dead nodes. Results uncover that Adaptive MCFL has as preferred standpoint over different methodologies. Table 4.2 describes the various applications in WSNs using neural networks, fuzzy logic, and a genetic algorithm.

The very robust WSNs work under challenging conditions, for example, keeping up message inertness and the dependability of information transmission and amplifying the battery life of sensor nodes. The point of this investigation is to propose a fluffy rationale calculation for comprehending these issues, which are hard to address with customary systems. The purpose is to take up a fuzzy logic method to utilize energy consumption and minimize packet drops [36]. This demonstration showed how fuzzy logic approach can be utilized to handle this particular communication issue with insignificant computational technique In this specific situation, the use of a fuzzy logic in the carrier sense multiple access with collision avoidance (CSMA/CA) system has been accomplished through filling the line length

TABLE 4.2

Description of Various Applications in WSN Using Neural Network, Fuzzy Logic, and a Genetic Algorithm

Sl. No	Literature	Year	Application area of Soft Computing in WSNs
1	Nauman Ahad et al. [1]	2016	Neural network modeling in WSNs
2	L. Muduli et al. [3]	2018	Fire control in coal mines using a fuzzy approach
3	S. Amri et al. [33]	2017	Node localization in WSNs using fuzzy logic
4	Nasim et al. [7]	2017	Fuzzy topology discovery in SDN-WSN
5	M. Mirzaie et al. [35]	2017	Multi-clustering algorithm using fuzzy method
6	I. Bouazzi et al. [36]	2017	Fuzzy controller in queuing prioritization
7	E. K. Eldhose et al. [4]	2016	Cluster node aggregation using neural networks
8	G. Sharma et al. [2]	2017	Range-free localization issue in 3D-WSN using a genetic algorithm
9	R.R. Swain et al. [37]	2016	Fault diagnosis using fuzzy MLP
10	H. Rajesh et al. [8]	2013	Data aggregation methods using soft computing

and the movement rate at every node. Through recreations, this demonstrates the proposed method has a superior execution as far as vitality utilization contrasted with the essential usage of CSMA/CA.

Article [37] presents a fault-finding convention for WSNs in light of a neural network approach. A PSO-founded fuzzy multilayer perceptron is used in the fault detection and classification phase of the protocol. The proposed convention considers the composite fault model, for example, hard perpetual, soft lasting, irregular, and transient fault. The execution of the proposed methodology is assessed by utilizing nonspecific parameters, for example, recognition precision, false caution rate, and a false positive rate. The recreation is done by standard network simulator NS-2.35 [38,39] and the execution is contrasted and the current fault analysis conventions. The outcome demonstrates that the proposed convention performs predominant than the current conventions.

4.3.4 Evolutionary Computation and Application in WSNs

Adding cognizance to existing WSNs with a subjective systems administration approach, which manages utilizing comprehension to the whole system convention stack and accomplish end-to-true objectives, achieves numerous advantages. However, intellectual systems administration might be mistaken for a subjective radio or cross-layer plan [40,41], therefore one might consider an alternate idea; psychological radios that apply comprehension just at the physical layer to defeat the issue of range shortage. Cross-layer configuration ordinarily concentrates on connecting no less than two non-back-to-back particular layers, to accomplish a specific objective. In fact, one might say that the intellectual radio and the cross-layer configuration are two successful strategies in psychological system administration [42,43]. To the best of our insight, the greater part of the current inquires about the cognitive wireless sensor networks (CWSNs) have concentrated on range designation and impedance diminishment in the physical layer. The authors in [27] proposed another thinking and learning model for CWSNs, in which right off the bat, a group of learning automata are utilized to build a Bayesian network (BN) model of the parameters of the system convention stack, and afterward the constructed BN is utilized to tune the controllable parameters [44,45]. The BN speaks to the reliance connections between the parameters of the system convention stack, and BN-based thinking is a productive apparatus for cross-layer optimization, so as to boost the apparent system execution. Reenactments have been done to assess the execution of the proposed method. The consequences of the recreations demonstrate that the proposed display progressively adds cognizance to a WSN and enhances the execution of the corresponding arrangement.

The authors in [2] presented an appropriated go free node confinement calculation for three-dimensional WSN, named as 3D-GAIDV (3D genetic algorithm based improved distance vector) hop. In this proposed algorithm, the normal hop size of anchor node is changed by refreshing a remedy factor and the adjusted jump measure is additionally streamlined by line seek calculation. The idea of coplanarity is acquainted with fewer area mistakes caused by the stay nodes, which are coplanar. The restriction exactness is additionally enhanced by applying hereditary calculation (genetic algorithm). To enhance the situating scope of the network, those objective nodes, which have been confined effectively in the first round of the restriction procedure, have moved up to associate stay node. Every one of the procedures and counts are completed at target node level, which makes the proposed calculation energy productive. Reenactment demonstrates that the proposed calculation outflanks the comparable sort of existing calculations.

Wireless sensor networks with steady sink nodes frequently experience the ill effects of problem areas, since sensor nodes near the sink ordinarily have more movement weight to forward amid the transmission process [46,47]. Using a mobile sink is a powerful method to improve the network execution, energy productivity, network lifetime, and dormancy, and so on [48,49]. The authors in [31] proposed a molecule swarm optimization-based clustering algorithm with a mobile sink for a remote sensor network. In that algorithm, the virtual clustering method was performed amid the steering process which influences utilization of the molecule to the swarm optimization algorithm. The leftover energy and position of the nodes were the essential parameters to choose the cluster head [50,51]. The control technique for the mobile sink to gather information from the cluster head was outlined in detail. Broad recreation comes about demonstrating that the energy utilization was abundantly diminished, the network lifetime was drawn out, and the transmission delay was decreased in our proposed steering algorithm than some other mainstream directing algorithms.

A distributed k-means algorithm [52] and a distributed fuzzy c-means approach has been formed in [52] for WSNs where every node is operational with sensors. The basic topology of the WSN should be emphatically associated. The contract algorithm in the multiagent agreement hypothesis was used to trade the estimation data of the sensors in WSN. To get a quicker meeting speed and a higher plausibility of having the worldwide ideal, a conveyed k-means++ methodology was first proposed to locate the underlying cancroids previously executing the disseminated k-means methodology and the appropriated fuzzy c-means methodology. The proposed dispersed k-means methodology was equipped for apportioning the information viewed by the nodes into measure-subordinate gatherings which have little in-gathering and extensive out-aggregate separations, while the proposed circulated fuzzy c-means methodology is fit for parceling the information saw by the nodes into various measure-subordinate gatherings with degrees of participation going from zero to one. Reenactment demonstrated that the proposed appropriated methodologies can accomplish nearly an indistinguishable outcome from that given by the brought together bunching methodology.

4.4 The Projected New SCSC Approach Based on Swarm Intelligence

After getting inspiration from a detailed survey on various issues related to smart applications of soft computing techniques from the above analysis, Chapter 4 presents an improved healthcare system called SCSC (soft computing for smart care) for the healthcare systems in a smart city perspective. Figure 4.1 shows the basic block diagram of the proposed approach.

The basic objective of the proposed SCSC approach is to provide safe travel to the patients in a most advanced VANET (vehicular ad hoc networks) [53] while traveling towards the healthcare unit and hospitals. Therefore, in a VANET [53], a vehicular communication based automated and intelligent computing system [54,55] has been developed that works fine within the range of SCSC controller. It uses a soft computing-based swarm intelligence logic [56,57] at the sensor device which controls the basic communication in the proposed smart healthcare system. The vehicles used in this system are also smart ambulances that are designed with smart sensors [58,59] equipped within them. The smart ambulance in the proposed system uses smart PCB (printed circuit board) attached to the smart vehicles with sensors reliable for providing different signals to the main controller. While ambulances transit in a smart range of SCSC-enabled campus, it can send four types of signal to the controller to get the corresponding healthcare and security service.

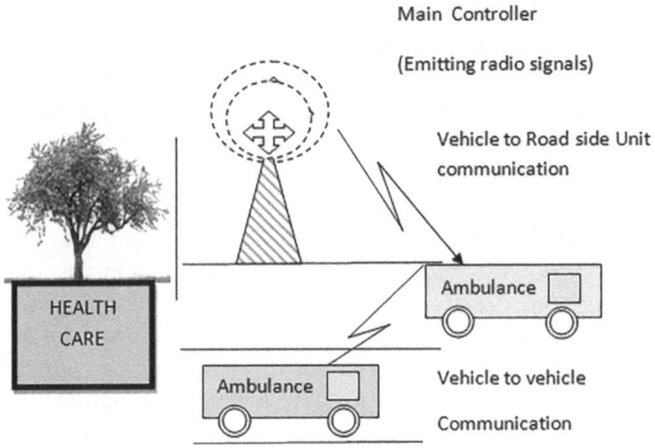

FIGURE 4.1
Block diagram of the proposed approach.

4.4.1 Safe Healthcare Mechanisms

This section explains the safety mechanism in intelligent and safe SCSC system which has been proposed in the aforementioned research. As shown in Figure 4.1, a main controller has been deployed at every major healthcare unit or roadside assistance centers. A SCSC controller unit is an integrated collection of the following components: a smart reader, for receiving input signals both at the hospital, roadside units, and ambulances; an antenna, as shown in Figure 4.1, and the periodic signals are emitted through the antenna; a controller that operates the built-in logic of nanobot sensor; an actuator, for input and output signals in real-time applications; a software system that executes at the controller as an operating system; and a communication infrastructure, which is a wireless network that connects all the components.

In the proposed approach, swarm intelligence techniques of soft computing methods have been used in a security module [60,61] using nanorobots. This logic senses the four basic security alert messages by the ambulance and services the patients in emergency [62,63]. It performs SAFE (SoS, attack, fuel shortage, and emergency) activities in smart-registered ambulances during travel in a smart VANET-based [64] SCSC zone. The SAFE approach handles the following safety and security mechanisms in smart ambulances:

SOS: This option is used for enabling "Save Our Soul" option. This can be triggered by the smart vehicle driver or passenger during any major problem. The roadside device automatically senses the alerts and alarms a signal at the human-assistance center to track the vehicle and provide emergency service.

Attack: Used to activate a button during any external attack on the vehicle, such as robbery and theft.

Fuel: In case of sudden fuel deficiency, the vehicle can be tracked and immediate assistance can be provided after sensing activation of this button in a smart vehicle.

Emergency: This emergency button can be used for either serious health hazard of driver or passenger during travel, any crime that has suddenly occurred inside the vehicle or, in the case of sudden accidents, this button can be pressed to send the emergency signal [65,66] to the SCSC controller and immediate assistance can be provided after informing the local officials. Using intelligent nanorobots both at the controller side and the ambulance side, an ambulance in the system always gets information [67,68] regarding the congestion status of the surrounding traffic. This communication takes place according to the V2V (vehicle-to-vehicle communication of VANET) [53] concept so that when it senses heavy traffic at the current traffic scenario, it can easily divert an ambulance's route towards the alternate path, hence saving time.

4.4.2 Smart Healthcare of Patients in Ambulance

Every patient and their relatives are safe in the system as the smart ambulance considered in this model carries the patient's personal information in its microcontroller chip, embedded in the vehicle's PCB (printed circuit board), so after entering the VANET [53] zone of SCSC, the ambulances are controlled and monitored by the controller. When the SCSC controller senses traffic jam conditions, it starts assisting the ambulances at their arrival point by diverting their path.

As shown in the Figure 4.2, along with other identifying information, the ambulance contains four basic security fields for communication with the main controller. During urgent situations, any one of the buttons in the ambulance will be activated either by the driver or by attendant of the patient, whoever is present. This is constantly being tracked by the main controller at the city points or roadside, using the V2R (vehicle-to-roadside unit) concept [69,70]. Upon receiving any vigilant message, immediate support and action is taken by the system to help the patients.

The presented approach supports security assurance using swarm intelligence as a soft computing technique for implementing improved ambulance driving control and safety features with an integrated approach of solving general traffic-related issues during any ambulance traveling in high volume traffic. Nanorobots [69] are used in this system, which work collaboratively as agents in the proposed swarm intelligence-based approach.

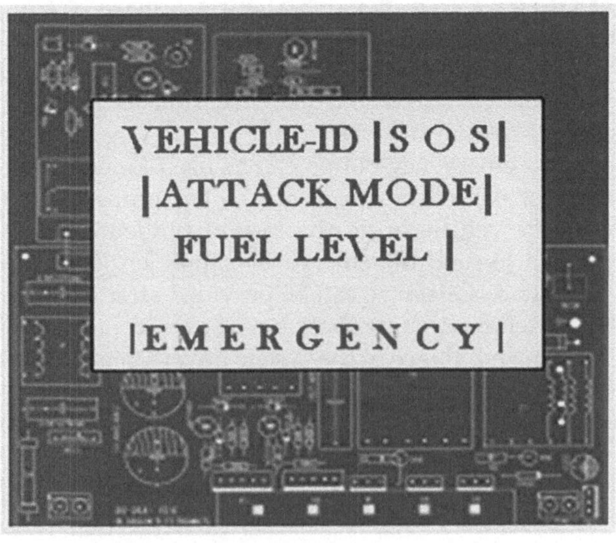

FIGURE 4.2
PCB of ambulance in proposed system.

4.5 Conclusion

Chapter 4 demonstrated a methodical survey that has been carried out on various soft computing applications in a very robust and efficient wireless sensor network. The advantages of using soft computing techniques in intelligence-based applications have been reviewed and analyzed. Many applications, their utilities, and experimental methodologies have been discussed, such as big data in an intelligent transport system, improved tourism applications, localization problems in WSNs, using ant bee colony optimization in network issues, and PSO in handover operations. Network design issues, such as WiMax network planning using a soft computing approach, optimization using neural networks, or cognitive WSN based on a Bayesian network have been analyzed. Various security issues like anomaly detection in WSNs and network size determination using very talented machine learning approaches have been studied. Basically, four types of excellent soft computing strategies, such as swarm intelligence, fuzzy logic, neural networks, and genetic algorithms, have a vast range of applications in wireless sensor networks. After an analytical study, it has been observed that few technical aspects of soft computing use learning automata and fuzzy logic for cognitive WSN network-based industrial applications. Some of the applications are applied in WSN cluster management, including multi-clustering algorithms and prioritization of queuing models using fuzzy controllers.

Finally, Chapter 4 gradually comes to an end with the conclusion that soft computing and artificial intelligence technologies have emerged as a complete solution for the emerging problems due to increasing network users and application-based issues in WSNs. Many PSO–based clustering algorithms have solved several problems in WSNs related to localization and fault diagnosis. This survey on soft computing techniques has also included the prevention and restoration of network failures in WSNs, error diagnosis in network configuration, and security mechanism in WSNs. As a part of our upcoming research, we have planned to extend this survey chapter to include more current application-oriented issues related to recent dynamic networks, such as cloud computing, fog computing, and ubiquitous computing. At the end of the analysis part, a novel and improved healthcare system has been proposed using the swarm intelligence logic of soft computing techniques. The proposed approach provides security assurance using soft computing techniques for implementing improved driving control and safety features with an integrated approach of solving general traffic-related issues during any ambulance traveling in a high-volume traffic gateway. Nanorobots are used in this system which work collaboratively as agents in the proposed swarm intelligence–based approach. In future work, we have planned to perform simulation in a real healthcare unit to ensure that it has better performance.

References

1. Nauman Ahad, Junaid Qadir, Nasir Ahsan. Neural networks in wireless networks: Techniques, applications and guidelines. *Journal of Network and Computer Applications* Elsevier, Volume 68, 2016, 1–27, ISSN 1084-8045.
2. Gaurav Sharma, Ashok Kumar. Improved range-free localization for three-dimensional wireless sensor networks using genetic algorithm. *Computers & Electrical Engineering* Elsevier, Available online 29 December 2017, ISSN 0045-7906.
3. Lalatendu Muduli, Prasanta K. Jana, Devi Prasad Mishra. Wireless sensor network based fire monitoring in underground coal mines: A fuzzy logic approach. *Process Safety and Environmental Protection* Elsevier, Volume 113, 2018, 435–447, ISSN 0957-5820.
4. Elby K. Eldhose, G. Jisha. Active Cluster Node Aggregation Scheme in Wireless Sensor Network Using Neural Network. *Procedia Technology* Elsevier, Volume 24, 2016, 1603–1608, ISSN 2212-0173.
5. Shidrokh Goudarzi, Wan Haslina Hassan, Mohammad Hossein Anisi, Ahmad Soleymani, Mehdi Sookhak, Muhammad Khurram Khan, Aisha-Hassan Abdalla Hashim, Mahdi Zareei. ABC-PSO for vertical handover in heterogeneous wireless networks. *Neurocomputing* Elsevier, Volume 256, 2017, 63–81, ISSN 0925-2312, https://doi.org/10.1016/j.neucom.2016.08.136.
6. P. P. Patel and R. H. Jhaveri. Soft computing techniques to address various issues in wireless sensor networks: A survey. *2016 International Conference on Computing, Communication and Automation (ICCCA)*, Noida, 2016, 399–404.
7. Nasim Abdolmaleki, Mahmood Ahmadi, Hadi Tabatabaee Malazi, Sebastiano Milardo. Fuzzy topology discovery protocol for SDN-based wireless sensor networks. *Simulation Modelling Practice and Theory* Elsevier, Volume 79, 2017, 54–68, ISSN 1569-190X.
8. Hevin Rajesh Dhasian, Paramasivan Balasubramanian. Survey of data aggregation techniques using soft computing in wireless sensor networks. *IET Information Security* IET Journals & Magazines, Volume 7, 336–342, 2013.
9. Chakchai So-In, Songyut Permpol, Kanokmon Rujirakul. Soft computing-based localizations in wireless sensor networks. *Pervasive and Mobile Computing* Elsevier, Volume 29, 2016, 17–37, ISSN 1574-1192.
10. Tomić, J. A. McCann. A Survey of Potential Security Issues in Existing Wireless Sensor Network Protocols. *IEEE Internet of Things Journal* Volume 4, Number 6, 1910–1923, December 2017.
11. Mamata Rath Bibudhendu Pati. Communication Improvement and Traffic Control Based on V2I in Smart City Framework. *International Journal of Vehicular Telematics and Infotainment Systems* (IJVTIS), Volume 2, Number 1, 2018.
12. Mamata Rath. An Exhaustive Study and Analysis of Assorted Application and Challenges in Fog Computing and Emerging Ubiquitous Computing Technology. *International Journal of Applied Evolutionary Computation* (IJAEC), Volume 9, Issue 2, April–June 2018, ISSN: 1942-3594; EISSN: 1942-3608; IGI-Global, USA, www.igi-global.com/ijaec.

13. R. R. Yager, A. J. Grichnik, R. L. Yager. A Soft Computing Approach to Controlling Emissions Under Imperfect Sensors. *IEEE Transactions on Systems, Man, and Cybernetics: Systems*. Volume 44, Number 6, 687–691, June 2014.
14. Mamata Rath, George S Oreku (Tanzania). Security Issues in Mobile Devices and Mobile Adhoc Networks. *Mobile Technologies and Socio-Economic Development in Emerging Nations* IGI Global 2018, 80, ISBN 152254030X, DOI-10.4018/978-1-5225-4029-8.ch009.
15. Mamata Rath, Jhum Swain, Bibudhendu Pati, Binod Kumar Pattanayak. Attacks and Control in MANET. *Handbook of Research on Network Forensics and Analysis Techniques*. IGI Global, Copyright: © 2018 |19–37, DOI: 10.4018/978-1-5225-4100-4.ch002.
16. S. Akila, R. Venkatesan. A Fuzzy Based Energy-aware Clustering Architecture for Cooperative Communication in WSN. *Computer Journal* Volume 59, Number 10, 1551–1562, October 2016.
17. Mamata Rath. Effective Routing in Mobile Ad-hoc Networks With Power and End-to-End Delay Optimization: Well Matched With Modern Digital IoT Technology Attacks and Control in MANET. *Advances in Data Communications and Networking for Digital Business Transformation* IGI Global, Copyright: © 2018 | 206–243, DOI: 10.4018/978-1-5225-5323-6.ch007.
18. Mamata Rath. A Multidimensional Review on Communication Issues in MANET and VANET. *Journal of Networking Computer Security and Engineering* Volume 3, Number 1, 2018.
19. S. Kumar. Compartmental Modeling of Opportunistic Signals for Energy Efficient Optimal Clustering in WSN. *IEEE Communications Letters* Volume 22, Number 1, 173–176, January 2018.
20. Bhatt, C., Dey, N., Ashour, A. S. (Eds.) *Internet of Things and Big Data Technologies for Next Generation Healthcare* 2017.
21. Rath, Mamata. An Analytical Study of Security and Challenging Issues in Social Networking as an Emerging Connected Technology (April 20, 2018). *Proceedings of 3rd International Conference on Internet of Things and Connected Technologies* (ICIoTCT), 2018 held at Malaviya National Institute of Technology, Jaipur (India) on March 26–27, 2018. Available at Elsevier, SSRN: https://ssrn.com/abstract=3166509.
22. Mamata Rath, Binod Kumar Pattanayak. A Novel Load Balanced Approach for Controlling Roadway Traffic in Smart City, 1st International Conference on New Frontiers in Engineering, Science & Technology, New Delhi, India, January 8–12, 2018.
23. Chao Wang, Xi Li, Xuehai Zhou, Aili Wang, Nadia Nedjah. Soft computing in big data intelligent transportation systems. *Applied Soft Computing* Elsevier, Volume 38, 2016, 1099–1108, ISSN 1568-4946, https://doi.org/10.1016/j.asoc.2015.06.006.
24. Thirumalaisamy Ragunathan, Sudheer Kumar Battula, Jorika Vedika, V. Anitha, T. Tarun, M. Shiva Prasad, M. Uma Kalyani. ITTS: Intelligent Transport and Tourism System. *Procedia Computer Science* Elsevier, Volume 50, 2015, Pages 191–196, ISSN 1877-0509, https://doi.org/10.1016/j.procs.2015.04.089.
25. Sadik Kamel Gharghan,Rosdiadee Nordin, Mahamod Ismail. A Wireless Sensor Network with Soft Computing Localization Techniques for Track Cycling Applications. *Sensors* 2016, Volume 16, Number 8, 1043; doi:10.3390/s16081043.

26. Le Hoang Son, Pham Huy Thong. Soft computing methods for WiMax Network Planning on 3D Geographical Information Systems. *Journal of Computer and System Sciences* Elsevier, Volume 83, Issue 1, 2017, 159–179, ISSN 0022-0000, https://doi.org/10.1016/j.jcss.2016.06.009.
27. S. Gheisari, M.R. Meybodi. A new reasoning and learning model for Cognitive Wireless Sensor Networks based on Bayesian networks and learning automata cooperation. *Computer Networks* Elsevier, Volume 124, 2017, 11–26, ISSN 1389-1286, https://doi.org/10.1016/j.comnet.2017.05.031.
28. Girik Pachauri, Sandeep Sharma. Anomaly Detection in Medical Wireless Sensor Networks using Machine Learning Algorithms. *Procedia Computer Science* Elsevier, Volume 70, 2015, 325–333, ISSN 1877-0509.
29. Sidra Rashid, Usman Akram, Shoab A. Khan. WML: Wireless Sensor Network based Machine Learning for Leakage Detection and Size Estimation. *Procedia Computer Science* Elsevier, Volume 63, 2015, 171–176, ISSN 1877-0509.
30. S. Gheisari, M.R. Meybodi. LA-CWSN: A learning automata-based cognitive wireless sensor networks. *Computer Communications* Elsevier, Volume 94, 2016, 46–56, ISSN 0140-3664.
31. Jin Wang, Yiquan Cao, Bin Li, Hye-jin Kim, Sungyoung Lee. Particle swarm optimization based clustering algorithm with mobile sink for WSNs. *Future Generation Computer Systems* Elsevier, Volume 76, 2017, 452–457, ISSN 0167-739X.
32. S. J. Habib, P. N. Marimuthu, K. Sudha, Soft computing tool for restoring failures within wireless sensor networks. *2014 International Wireless Communications and Mobile Computing Conference (IWCMC)*, Nicosia, 2014, 86–91.
33. Saber Amri, Fekher Khelifi, Abbas Bradai, Abderrezak Rachedi, Med Lassaad Kaddachi, Mohamed Atri. A new fuzzy logic based node localization mechanism for Wireless Sensor Networks. *Future Generation Computer Systems* Elsevier, 2017 (In Press).
34. Energy Competent Routing Protocol Design in MANET with Real time Application Provision. *International Journal of Business Data Communications and Networking* IGI Global, Volume 11, Number 1, 50–60, January–March 2015.
35. Mostafa Mirzaie, Sayyed Majid Mazinani. Adaptive MCFL: An adaptive multi-clustering algorithm using fuzzy logic in wireless sensor network. *Computer Communications* Elsevier, Volume 111, 2017, 56–67, ISSN 0140-3664.
36. Imen Bouazzi, Jamila Bhar, Mohamed Atri. Priority-based queuing and transmission rate management using a fuzzy logic controller in WSNs. *ICT Express*, Elsevier, Volume 3, Issue 2, 2017, 101–105, ISSN 2405-9595.
37. R. R. Swain, P. M. Khilar. A fuzzy MLP approach for fault diagnosis in wireless sensor networks. *2016 IEEE Region 10 Conference (TENCON)*, Singapore, 2016, 3183–3188, doi: 10.1109/TENCON.2016.7848637.
38. Mamata Rath, Jhum Swain, Bibudhendu Pati, Binod Kumar Pattanayak. Attacks and Control in MANET. *Handbook of Research on Network Forensics and Analysis Techniques*. IGI Global, Copyright: © 2018 19–37, DOI: 10.4018/978-1-5225-4100-4.ch00.
39. Mamata Rath, George S Oreku. Security Issues in Mobile Devices and Mobile Adhoc Networks. *Mobile Technologies and Socio-Economic Development in Emerging Nations*. IGI Global 2018, 80, ISBN 152254030X.
40. Mamata Rath. A Multidimensional Review on Communication Issues in MANET and VANET. *Journal of Networking Computer Security and Engineering* Volume 3, Number 1, 2018.

41. Mamata Rath, Bibudhendu Pati. Communication Improvement and Traffic Control Based on V2I in Smart City Framework. *International Journal of Vehicular Telematics and Infotainment Systems* (IJVTIS), Volume 2, Number 1, 2018.
42. M. Rath, M. R. Panda. MAQ system development in mobile ad-hoc networks using mobile agents. IEEE 2nd International Conference on Contemporary Computing and Informatics (IC3I), Noida, 794–798. Online on January, 2017.
43. Rath M., Pattanayak B.K. SCICS: A Soft Computing Based Intelligent Communication System in VANET. Venkataramani G., Sankaranarayanan K., Mukherjee S., Arputharaj K., Sankara Narayanan S. (eds) *Smart Secure Systems – IoT and Analytics Perspective*. ICIIT 2017. *Communications in Computer and Information Science* Volume 808. Springer, Singapore.
44. Rath M., Pattanayak B.K. Monitoring of QoS in MANET Based Real Time Applications. Satapathy S., Joshi A. (eds) *Information and Communication Technology for Intelligent Systems* (ICTIS 2017) - Volume 2. ICTIS 2017. *Smart Innovation, Systems and Technologies* Volume 84, 579–586, Springer, Cham.
45. M. Rath, B. Pati, B. K. Pattanayak. Cross layer based QoS platform for multimedia transmission in MANET. 2017 11th International Conference on Intelligent Systems and Control (ISCO), Coimbatore, 2017, 402–407.
46. Rath M., Pattanayak B.K., Pati B. Energetic Routing Protocol Design for Real-time Transmission in Mobile Ad hoc Network. *Computing and Network Sustainability, Lecture Notes in Networks and Systems* Volume 12. Springer, Singapore, 2017.
47. Congestion Control Mechanism for Real Time Traffic in Mobile Adhoc Networks, Computer Communication, Networking and Internet Security. *Lecture Notes in Networks and Systems* Volume 5. Springer, 2017.
48. Rath M, Pattanayak B. MAQ: A Mobile Agent Based QoS Platform for MANETs. *International Journal of Business Data Communications and Networking*, IGI Global, Volume 13, Issue 1, 1–8, January 2018.
49. M. Rath, M.R. Panda. MAQ System Development in Mobile Ad-hoc Networks using Mobile Agents. IEEE Second International Conference on Contemporary Computing and Informatics (IC3I), Noida, 794–798, January 2017.
50. Rath M, Pattanayak B.K. SCICS: A Soft Computing Based Intelligent Communicxation System in VANET. *Communications in Computer and Information Science* Volume 808, Springer, 2018.
51. Energy Efficient MANET Protocol Using Cross Layer Design for Military Applications. *Defense Science Journal* Volume 66, Number 2, DRDO Publication, March 2016 (SCI Index, Web of Science, Scopus Indexed).
52. J. Qin, W. Fu, H. Gao, W. X. Zheng. Distributed K -Means Algorithm and Fuzzy C-Means Algorithm for Sensor Networks Based on Multi agent Consensus Theory. *IEEE Transactions on Cybernetics* Volume 47, Number 3, 772–783, March 2017.
53. H. J. F. Qiu, I. W. H. Ho, C. K. Tse, Y. Xie. A Methodology for Studying 802.11p VANET Broadcasting Performance With Practical Vehicle Distribution. *IEEE Transactions on Vehicular Technology* Volume 64, Number 10, 4756–4769, October 2015.
54. Comparative analysis of AODV routing protocols based on network performance parameters in Mobile Adhoc Networks. *Foundations and Frontiers in Computer, Communication and Electrical Engineering* 461–466, ISBN: 978-1-138-02877-7. May, 2016, CRC Press, Taylor & Francis, 2016.

55. Resource Reservation and Improved QoS for Real Time Applications in MANET. *Indian Journal of Science and Technology* Volume 9, Number 36, 2016. DOI: 10.17485/ijst/2016/v9i36/100910.
56. A Contemporary Survey and Analysis of Delay and Power Based Routing Protocols in MANET. *Journal of Engineering and Applied Sciences* 2016, Volume 11, Number 1, 536–540, January 2016.
57. QoS Satisfaction in MANET Based Real Time Applications. *International Journal of Control Theory and Applications* Volume 9, Issue 7, 3069–3083, 2016.
58. Inter-Layer Communication Based QoS Platform for Real Time Multimedia Applications in MANET. *Wireless Communications, Signal Processing and Networking* (IEEE WiSPNET) Chennai, India, 613–617, 2016.
59. Prioritisation of Security Measures at the Junction of MANET and IoT, Second International Conference on Information and Communication Technology for Competitive Strategies, ACM Publication, New York, USA, 2016. ISBN: 978-1-4503-3962-9.
60. Delay and power based network assessment of network layer protocols in MANET. 2015 International Conference on Control, Instrumentation, Communication and Computational Technologies, Kumaracoil, India, 682–686, 2015.
61. A Mobile Agent Based Intrusion Detection System Architecture For Mobile Ad Hoc Networks. *Journal of Computer Science* Volume 10, 970–975, 2014.
62. A methodical survey on real time applications in MANETS: Focussing On Key Issues International Conference on High Performance Computing and Applications, Bhubaneswar, India, 1–5, 22–24 December 2014.
63. Binh, H. T. T., Hanh, N. T. Dey, N. (2016). Improved cuckoo search and chaotic flower pollination optimization algorithm for maximizing area coverage in wireless sensor networks. *Neural Computing and Applications* 1–13.
64. S. Shamshirband et al. Soft-Computing Methodologies for Precipitation Estimation: A Case Study. *IEEE Journal of Selected Topics in Applied Earth Observations and Remote Sensing* Volume 8, Number 3, 1353–1358, March 2015.
65. Dey, Nilanjan, Amira S. Ashour, Fuqian Shi and R. Simon Sherratt. Wireless capsule gastrointestinal endoscopy: Direction-of-arrival estimation based localization survey. *IEEE Reviews in Biomedical Engineering* 10, 2–11, 2017.
66. Dey, N., Ashour, A. S., Shi, F., Fong, S. J., Sherratt, R. S. Developing residential wireless sensor networks for ECG healthcare monitoring. *IEEE Transactions on Consumer Electronics* Volume 63, Number 4, 442–449, 2017.
67. Dey, N., Hassanien, A. E., Bhatt, C., Ashour, A., Satapathy, S. C. (Eds.) *Internet of Things and Big Data Analytics Toward Next-Generation Intelligence*. Springer, 2018.
68. Elhayatmy, G., Dey, N., Ashour, A. S. Internet of Things Based Wireless Body Area Network in Healthcare. *Internet of Things and Big Data Analytics Toward Next-Generation Intelligence* 3–20. Springer, Cham, 2018.
69. Hsin-Te Wu, Gwo-Jiun Horng. Establishing an Intelligent Transportation System with a Network Security Mechanism in an Internet of Vehicle Environment. *IEEE Access* 2169–3536 (c) 2017 IEEE. Translations and Content Mining, DOI 10.1109/ACCESS.2017.2752420.
70. Yadav, P., Sharma, S., Tiwari, P., Dey, N., Ashour, A. S., Nguyen, G. N. A Modified Hybrid Structure for Next Generation Super High Speed Communication Using TDLTE and Wi-Max. *Internet of Things and Big Data Analytics Toward Next-Generation Intelligence*, 525–549. Springer, Cham, 2018.

5

Soft Computing Technique for Intrusion Detection System in Mobile Ad Hoc Networks

V. Bapuji, B. Manjula, and D. Srinivas Reddy

CONTENTS

5.1 Background and Driving Forces

Soft computing (SC) is the fusion of methodologies that were designed to model and enable solutions to real problems. These problems are typically associated with fuzzy, complex and dynamic systems with uncertain parameters. These systems are modeled to solve the real problems in computer applications.

The idea of soft computing was proposed by Dr. L. A. Zadeh [1] in 1981 and defined "soft computing in its latest incarnation as the fusion of fuzzy logic, evolutionary computing, neuro computing, probabilistic computing and genetic computing into one multidisciplinary system." The fundamental goal was to construct a novel artificial intelligence is known as computational intelligence. Soft computing is applied to smart machines to resolve non-linear and mathematically unmodeled system issues.

Different components have been proposed and more new techniques are still merging into soft computing [2,3]. The components include:

- Neural networks
- Fuzzy systems
- Evolutionary computation
  - Evolutionary algorithms
  - Harmony search
- Swarm intelligence
- Probability concepts like:
  - Bayesian networks
  - Chaos theory

5.2 Applications of Soft Computing

Soft computing provides two main advantages:

1. SC techniques are used to solve problems, even though it would be better if any mathematical model is available for solving non-linear problems.
   - Human features, such as:
     - Cognition
     - Recognition
     - Understanding
     - Learning

2. The following potential applications construct the intelligent automated systems, self-tuning spontaneous autonomous systems, etc.
   - Wireless sensor networks
   - Remote/air traffic control
   - Data/web Mining (cleaning and validation)
   - GPS
   - Image processing
   - Medical diagnosis and research
   - Target identification
   - Face/voice recognition
   - Man–machine communication
   - Animation and graphics
   - Nero fuzzy systems and fuzzy logic control

5.3 Properties of Soft Computing

The properties of soft computing were proposed [4] as:

- Learning from experimental data.
- Deriving the power of generalization from approximating to produce outputs from previously unseen inputs by using outputs from previously learned inputs.
- The generalization is usually in a high-dimensional space.

The literature review on this particular topic is discussed clearly and very detailed.

5.4 Intrusion Detection in MANETs

Providing security in mobile ad hoc networks is of prime concern due to the need for providing protected communication between mobile nodes in a hostile environment. Intrusion attempts any set of procedures to compromise the integrity, confidentiality or availability of a resource [5]. The detection of such kinds of intrusion is called an intrusion detection system (IDS). The IDS detects potential violations of a security policy by tracking system activities and responding to attacks that are intrusive.

IDS is a crucial part of a security system, because intrusion techniques are not always adequate and new intrusions commonly emerge. Intrusion detection systems detect possible violations of a security policy by observing system activity and responses. Once any attack is identified in the network through an alarm, the damage can be prevented or minimized.

Chapter 5 discusses the various methods to detect intrusion in mobile ad hoc networks. State-of-the-art solutions, such as soft computing techniques, are inspired by the natural expansion of grammatical evolution to detect attacks and synthesize detection programs on mobile ad hoc networks. The performance of these programs is evaluated on different types of networks with different mobility and traffic patterns to show their effect on the intrusion detection of ad hoc on demand vector (AODV) attacks.

5.5 AODV Routing Protocol

AODV is a reactive routing protocol that is designed for wireless and mobile ad hoc networks [6–8]. In AODV, the networks are silent until the connections are established and routes to destination are only on demand by source nodes with unicast and multicast routing. The destination sequence numbers are employed to specify how fresh a route is used to grant loop freedom [9].

Every time a node sends the request packet (REQ) to its destination. If it is not fresh enough, then it broadcasts route request (RREQ) messages to its neighbors for an alternate route path. Each neighbor node that receives the broadcast message sets up a reverse route towards the discoverer of the RREQ, except it has a *fresh* one. The RREQ packet contains the hop count, sequence numbers, broadcast ID, source, and destination address.

When a destination node receives the RREQ (or an intermediate node that has a "fresh enough" route to the destination), it responses with an RREP (route reply). The variable information in RREP and RREQ is the "hop count", which is continuously increased whenever it is important.

In this case, RREQ is replied by any of its intermediate nodes; they also need to send a route reply packet to the destination. In this way, the bidirectional route path is set up. In this scenario, the node receives a new route (by an RREQ or by an RREP) and the node already has a route as *fresh* as the received one and the shortest one will be updated. The following are the various attacks on AODV routing protocols:

- Packet dropping
- Ad hoc flooding
- Route disruption

In Chapter 5, route disruption and ad hoc network flooding attacks are investigated: the results are promising and are well discussed.

5.6 Packet Dropping Attack

In mobile ad hoc networks, a node performs both as a terminal and routing function to form an infrastructure-less network. Therefore, a node is assumed to be helpful to other nodes to forward packets toward the accurate destination. Dropping attack means a node does not forward packets for other nodes and silently or intentionally drop the packets. In this attack, malicious nodes drop data packets not intended for it any given time of interval to interrupt the network connection. In MANETs, packets may be dropped for several reasons such as collisions, channel errors and buffer overflows.

5.7 Flooding Attack

In mobile ad hoc networks, the topology changes frequently due to mobility. Furthermore, link breakage is very common in wireless networks. The ad hoc flooding attack is a denial-of-service (DoS) attack. In this attack, the source node acts as a malicious node and generates a large number of RREQs or RREP packets until the network is saturated.

If any malicious node intentionally disrupts the network operations or other node activity in the network, then the malicious node initiates the route discovery process. It tries to find the path for existing, nonexisting, or unavailable nodes in the network. For this reason, malicious nodes floods the RREQ packet an infinite number of times. The other participating nodes in a network become unable to receive or handle other packets. Due to the flooding of RREQ, the intermediate node cannot concentrate forwarding and other activities.

It is very difficult to distinguish an ad hoc flooding attack from a benign behavior of a network due to the mobility of nodes [10]. This attack can seriously degrade the performance of reactive routing protocols and affect a node in following ways:

- Performance in buffer
- Performance in wireless interface
- Performance in RREQ packets
- Performance in lifetime of MANETs

5.8 Route Disruption Attack

In this attack, without receiving a route request message, the attacker sends route reply messages to the victim node. There is no method to check the route request–reply flow in AODV. Even though the nodes are overhearing the route reply messages, they can update the routing table if the message carries fresh route information.

Mostly, in random destination nodes, the attacker selects one of the neighbor nodes as a victim, and will not send route replies to the random destination nodes. Here, the attacker knows about the active routes of the victim node through the routing control packets, which are broadcasted even though the attacker is a neighbor to the victim. Since the RREP message sent by the attacker is fresh, the routing table is updated by the victim node with the routes that the attacker advertises. Thus, the attacker is able to interrupt the victim's active routes.

5.9 Proposed Work

5.9.1 Grammatical Evolution and Algorithm

In this procedure, programs are generated in an arbitrary language by evolving programs written in a BNF grammar [11–13]. In this algorithm, a problem is defined with a *fitness function* and *grammar* (Figure 5.1). Table 5.1 shows the code and notation of a grammar that is used to evolve the programs and detect the flooding and route disruption attacks of MANETs.

TABLE 5.1

Backus–Naur Form Grammar

<table>
<tr><th>S =</th><th><notation></th></tr>
<tr><td><code></td><td>:: = if (<condition>) {raise_alarm()}</td></tr>
<tr><td><condition></td><td>:: = <condition> <set-op> <condition> | <expr1> <relop> <expr2></td></tr>
<tr><td><expression></td><td>:: = <expr> <op> <expr> | (<expr> <op> <expr>) | <pre-op1> (<expr>) | pre-op2) (<expr>) | <var></td></tr>
<tr><td><operators></td><td>:: = + | - | / | x</td></tr>
<tr><td><pre-op1></td><td>:: = sin | cos | log | ln | sqrt | abs |</td></tr>
<tr><td><pre-op2></td><td>:: = maximum | minimum | power | percent</td></tr>
<tr><td><rel-op></td><td>:: = < | ≤ | > | ≥ | == | !=</td></tr>
<tr><td><set-op></td><td>:: = and | or | not</td></tr>
<tr><td><var></td><td>:: = Tables 5.3 and 5.4 represent features</td></tr>
</table>

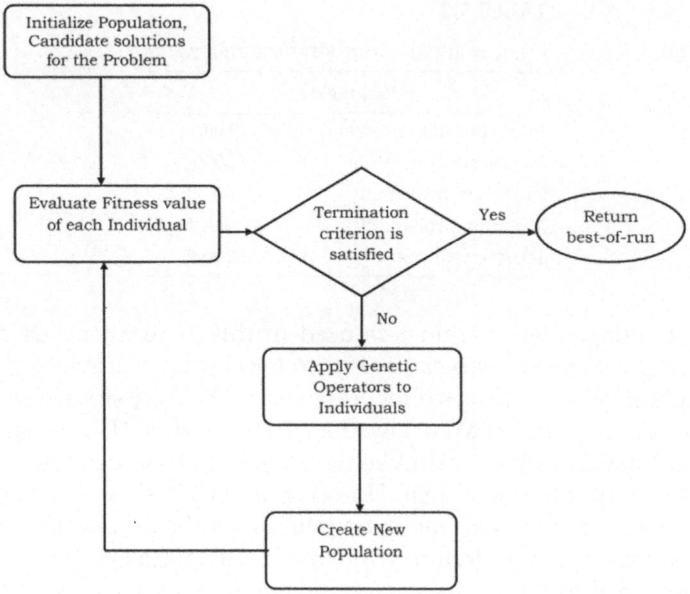

FIGURE 5.1
Grammatical evolution algorithm. (From V. Bapuji et al., *Networks and Complex Systems*, ISSN 2224-610X (Paper) ISSN 2225-0603 (Online), 24–31, Vol. 2, No. 4, http://www.iiste.org/Journals/index.php/NCS/article/view/2374/2373, 2012.)

5.9.2 Pseudocode of the Grammatical Evolution

5.9.2.1 Initialize (Population)

while *(termination criterion is not true)* ***do***
execute (evaluate each individual fitness value)
apply genetic_ operators (selection, crossover, mutation, etc.)
create (new_ population)
terminate ***while loop***
return *(best-of-run each individual)*
end

5.10 GE Parameters

In grammatical evolution, a problem is defined with a grammar and a fitness function. The following GE parameters are used for experiments (as shown in Table 5.2) to detect ad hoc flooding and route disruption attacks of MANETs.

TABLE 5.2

Grammatical Evolution Parameters

Parameters	
Initial population size	100
Maximum generations	2000
Crossover probability	0.1 to 0.9
Mutation probability	0.01 to 0.5
Tournament size	2 to 9

The five independent parameters used in this experiment are *crossover, mutation, population size, tournament size,* and *maximum number of generations.*

The highest value (0.9) is used for crossover probability since leaving some part of the old population to survive the next generation. This is assumed to be good. Mutation helps to avoid being trapped in local extremes by introducing diversity into the system. However, mutation should be used sparingly (the use of a very high mutation rate causes the process to degenerate into a random search). Therefore, a maximum rate of 0.5 is used.

In Chapter 5, the high mutation rates did not give good results. The parameter for the termination criteria of the grammatical evolution algorithm, namely generations, depends upon the parameter of *population size.*

Crossover probability shows how likely this operator will be performed on the individuals selected for mating (allowing the current candidates to proceed the next generation prior to any mutation).

The probabilities of mutation represent how likely a part of an individual will be mutated. The reproduction probability shows that an individual will be copied without any modification in the new generation.

A group of individuals are selected randomly in a tournament selection. From that population, and the fittest individuals from this group are selected as parents. *Tournament size* defines the maximum number of individuals in this group.

The population size is typically constant most of the time in grammatical evolution. Subsequently, the individuals who will survive in the next generation must be selected. A choice is made among the current population and the new individuals generated by variation operators. This choice is based on the fitness value.

5.11 Feature Selection

The *Features* are an input of the proposed system. The features are classified into two groups:

Mobility Features: These features help the mobility model of a node or the network, shown in Table 5.3.

TABLE 5.3

Mobility-Related Features

Mobility-Related Features	Description
neighbors	#neighbors
added_neighbors	#neighbors added
remd_neighbors	#neighbors removed
ative_routes	#active routes
repd_routes	#routes are under repair
invd_routes	#invalidated routes
addedrts_disc	#added routes by route discovery mechanism
addedrtes_notice	#added routes by overhearing
updated_routes	#updated routes (modifying hop count, sequence number)
added_repaired routes	#added routes under repair
invroutes_timeout	#invalidated routes due to expiry
invroutes_other	#invalidated routes due to other reasons
avg_hopcount	average hop counts of active routes

#number of

Packet Features: Table 5.4 shows the packet-related information about the frequency of the routing protocol control packets (RREQ, RREP, and RERR) and whatever is sent, received, or forwarded in a time interval and updated.

In grammatical evolution, the fitness function is the most important and is used to calculate the difference between the detection rate and the false positive rate.

$$\text{Fitness} = \text{Detection Rate} - \text{False Positive Rate} \tag{5.1}$$

The fitness function is based on the main metrics used to evaluate an intrusion detection system, such as the detection rate and false positive rate. This is used to estimate the effectiveness of a proposed system.

The detection rate is the ratio of correctly detected attacks to the total number of attacks on the network.

$$\text{Detection rate} = \frac{\text{Correctly detected attacks}}{\text{Total attacks}} \tag{5.2}$$

The false positive rate (FPR) represents the ratio of normal activities that are incorrectly identified as intrusions to the total of normal activities on the network.

$$\text{False positive rate} = \frac{\text{Normal activities incorrectly detected as attacks}}{\text{Total attacks}} \tag{5.3}$$

TABLE 5.4

Packet-Related Features

Packet-Related Features	Description
recv_rreqPs	#received route request packets destined to this node
recvF_rreqPs	#received route request packets to be forwarded by this node
send_rreqPs	#broadcasted route request packets from this node
frwd_rreqPs	#forwarded route request packets from this node
recv_rrepPs	#received route reply packets destined to this node
recvF_rrepPs	#received route reply packets to be forwarded by this node
send_rrepPs	#initiated route reply packets from this node
frwd_rrepPs	#forwarded route reply packets from this node
recvB_rerrPs	#received broadcast route error packets (to be forwarded or not)
send_rerrPs	#broadcasted route error packets from this node
recv_aodvPs	#received total routing protocol packets
recvF_aodvPs	#received total routing protocol packets to be forwarded
send_aodvPs	#initiated total routing protocol packets from this node
frw_aodvPs	#forwarded total routing protocol packets by this node

#number of

5.12 Simulation Study

The evolved programs are evaluated on the networks simulated by Ns-2 [14] and the performance of grammatical evaluation is analyzed through different attacks and mobility patterns of the nodes.

The selection of relevant features is very important for any machine learning method. However, too many features might degrade the performance of the learning algorithm.

Chapter 5 mainly focuses on reducing the features from Tables 5.3 and 5.4 for better performance, using the genetic algorithm and the selection of right features for training. Hence, this technique is used to increase the performance of the proposed algorithm by using the relevant features.

When the numbers of nodes are selected, their mobility and packet related features are selected while simulating ad hoc flooding and route disruption attacks.

5.12.1 Ad Hoc Flooding Attack

- Send_rrepPs,frwd_aodvPs,updated_routes,frwd_rreqPs,added_repairedroutes.

5.12.2 Route Disruption Attack

- Invroutes_other,recv_aodvPs,send_aodvPs,send_rreqPs,repaired_routes,recvB_rerrPs.

The above features are considered for this study in particular for the following reasons:

- Ad hoc network routing behavior is very complex.
- Genetic algorithms have been used for the selection and reduction of features in many areas successfully [11,15,16].
- Only the relevant features are applied by using the proposed algorithm in order to increase the performance of detection of these attacks.

In this approach, each and every node moves from its present location to its random new location within the determined random speed/pause time limits. Table 5.5 shows the simulation parameters of the network.

As the network dimensions and parameters increase, the false positive rate decreases considerably, and this requires more TCP connections. Hence, the above moderate parameters are selected for simulations.

Different mobility levels and traffic loads are created under different scenarios. Seventy nodes are placed in a topology 1100 m × 500 m. The TCP traffic is used for communication. The maximum numbers of 20–30 TCP connections are used to simulate on different traffic loads. The nodes' maximum speed is set to 20 m/s and pause time among activities is 300, 600, and 900 seconds to simulate low, medium, and high mobility, respectively.

The AODV routing protocol selects a periodic *"Hello"* message that is used for local link connectivity. Simulations are performed for 5000 seconds for training data and 2000 seconds for testing data. The detection programs are evolved using the training data collected from a network under medium mobility with 30 TCP connections.

TABLE 5.5

Simulation Parameters

Network dimensions	1100 × 500 meters
Number of nodes	70
Packet traffic	20 and 30 connections using TCP
Speed	0–30 m/s, pause time 300, 600, and 900 seconds. low/medium/ high mobility, respectively
Protocol	AODV
Propagation model	250 m transmission range with two-ray ground propagation model
Connectivity	*"Hello"* message AODV periodic local link
Time	5000 secs (training) and 2000 secs (testing)

5.13 Experimental Results

The intrusion detection programs are distributed to each node in a network. Each node collects the relevant features at each interval time. The packet loss occurs in MANETs frequently due to congestion, mobility of nodes, and wireless link transmission errors. Thus, the features efficiently detect the attacks.

- In an ad hoc flooding attack, the nodes are flooded by route request messages to detect the attack.
- In a route disruption attack, the injured node is assumed to detect malicious change in routing table.

5.14 Ad Hoc Flooding Attack False Positive Rate

The evolved program's performance is analyzed on the network under different mobility and traffic patterns with 300, 600, and 900 seconds separately by collecting trained data from the best individuals of an ad hoc flooding attack, as shown in Table 5.6.

At different time intervals in this scenario, the attacker exploits the property of link-breakages and route discovery mechanisms by broadcasting lots of RREQ packets to arbitrarily selected nodes.

TABLE 5.6

Simulation Results of Ad Hoc Flooding Attack Mobility and Traffic Levels

	Ad Hoc Flooding Attack		
	False Positive Rate		
Mobility and Traffic Levels	**300 Secs**	**600 Secs**	**900 Secs**
Low mobility Low traffic	1.1	1.3	2.1
Low mobility Medium traffic	2.2	4	5.1
Medium mobility Low traffic	1.3	1.9	2.7
Medium mobility Medium traffic	2.1	3.9	4.9
High mobility Low traffic	1.4	2	2.8
High mobility Medium traffic	2.6	4	5.2

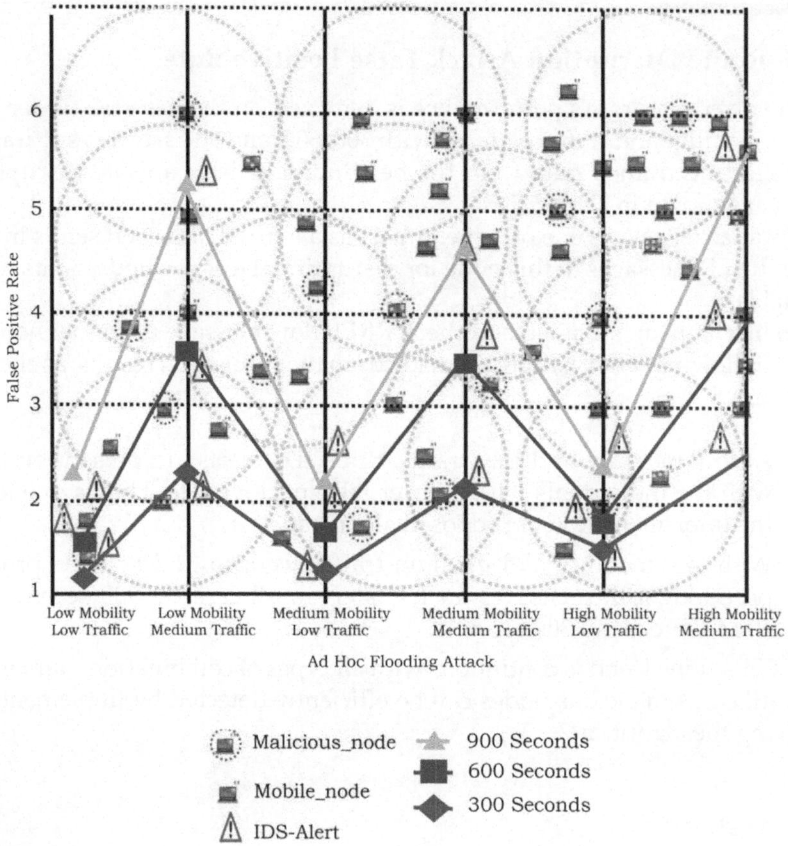

FIGURE 5.2
Performance variations of mobility and traffic in ad hoc flooding attack.

The following observations are made from the simulation results, as shown in Figure 5.2.

- At a low mobility and a low traffic scenario, three IDS alerts are present even with a low false positive rate. This is due to the proposed algorithm.
- With low mobility and medium traffic, we found that the proposed algorithm ensures an IDS alert in all cases at 900 seconds, 600 seconds, and 300 seconds.
- This same trend continues with all types of combinations, and in all cases malicious nodes can be efficiently detected by implementing the algorithm. We found that at every given time, at least one IDS alert is found.

5.15 Route Disruption Attack False Positive Rate

The evolved program's performance is analyzed on the network under different mobility and traffic patterns with 300, 600, and 900 seconds separately by collecting trained data from the best individuals of a route disruption attack, as shown in Table 5.7.

The attacker selects one of its neighbors as a injured node and sends maximum RREP messages to this node for disrupting the active nodes in its routing table.

No mechanism is available to check RREP flow message from that node.

The following observations are made from the simulation results, as shown in Figure 5.3:

- At low mobility and low traffic, three IDS alerts are present even with the individuals with a lower false positive rate. This is due to the implementation of proposed algorithm.
- With low mobility and medium traffic, we also find that the proposed algorithm ensures an IDS alert in all cases at 900 seconds, 600 seconds, and 300 seconds.
- This same trend is continuous with all types of combinations, and in all cases malicious nodes can be efficiently detected by implementing the algorithm.

TABLE 5.7
Simulation Results of Route Disruption Attack Mobility and Traffic Levels

	Route Disruption Attack		
	False Positive Rate		
Mobility and Traffic Levels	300 Secs	600 Secs	900 Secs
Low mobility Low traffic	1.2	1.4	2.1
Low mobility Medium traffic	3.4	4.4	5.6
Medium mobility Low traffic	1.3	2	2.5
Medium mobility Medium traffic	2.8	3.9	5.1
High mobility Low traffic	1.4	2	2.8
High mobility Medium traffic	3.2	3.8	5.8

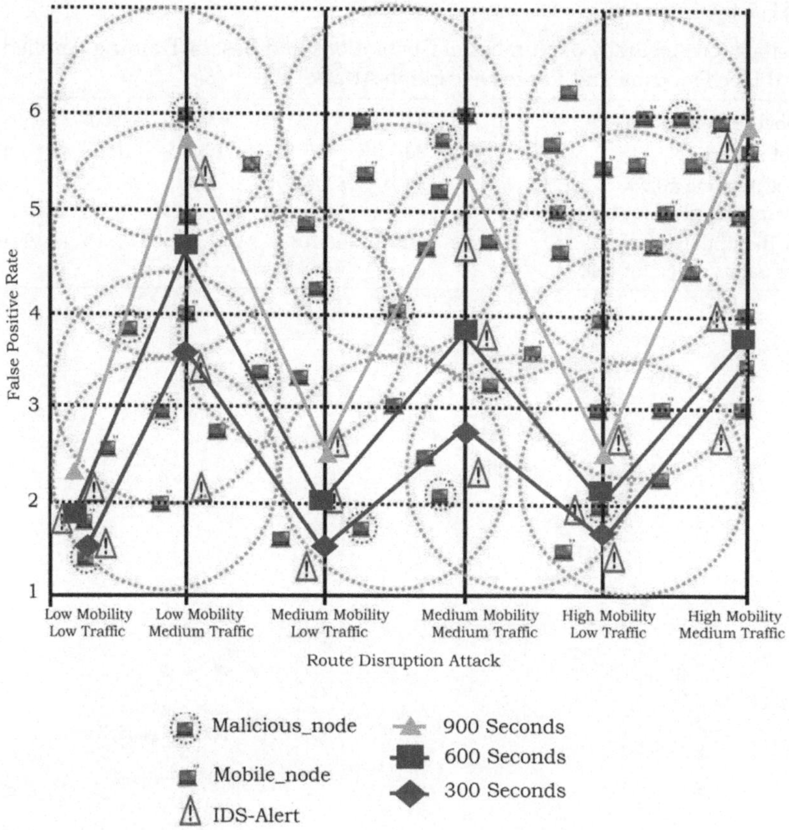

FIGURE 5.3
Performance variations of mobility and traffic in route disruption attack.

5.16 Classification Accuracy of Ad Hoc Flooding Attack and Route Disruption Attack

The classification accuracy of the ad hoc flooding and route disruption attack on the number of generations is shown in Table 5.8 and analyzed. Only the best trained generations are applied for test.

The following observations are made from the simulated results represented in Figure 5.4.

- The classification accuracy of the ad hoc flooding attack is 61.01% and the route disruption attack is 70.01% of best train applied for 100 generations. Because in both cases at an early stage, the fitness of the survival of generations is much lower, hence the accuracy is low.

TABLE 5.8

Classification Accuracy of Number of Generations and Best of Training Applied for Ad Hoc Flooding and Route Disruption Attack

Number of Generations	100	500	1000	1500	2000
Best of training	80	95	97	100	100
Ad hoc flooding attack (Best train applied for test)	61.01%	82.30%	94.23%	96.78%	99.88%
Route disruption attack (Best train applied for test)	70.01%	88.10%	95.12%	97.81%	99.98%

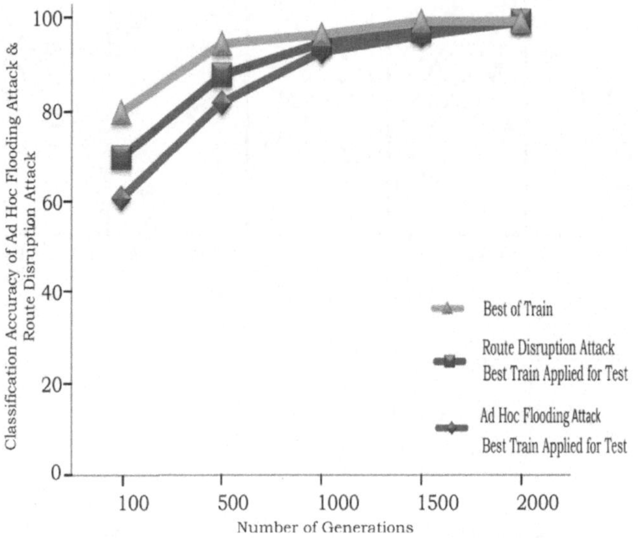

FIGURE 5.4
GE: Classification accuracy of ad hoc flooding and route disruption attack on number of generations.

From Figure 5.4 after comparison, this is concluded by applying the proposed algorithm: the route disruption attack has overall better performance than the ad hoc flooding attack.

It is observed that at 2000 generations, the maximum (high) fitness is achieved by applying the proposed algorithm, because the latest populations are generated iteratively by using selection, crossover, and mutation operators as on natural evolution. The classification accuracy of ad hoc flooding and route disruption attacks of best train is applied for tests are continued at 2000 generations.

TABLE 5.9

Ad Hoc Flooding and Route Disruption Attacks Grammatical Evolution Performance

		False Positive Rate			
Attack Type	**Detection Rate**	**Low**	**IDS Alert**	**High**	**IDS Alert**
Ad hoc flooding	99.88%	1.1%	Available	5.2%	Available
Route disruption	99.98%	1.2%	Available	5.8%	Available

5.17 Grammatical Evolution Performance of Ad Hoc Flooding and Route Disruption Attack

The grammatical evolution algorithm is run on the training data 10 times for each attack type and then evaluated on different mobility patterns. The best result out of 10 runs is evaluated on the simulated networks.

The experimental results show that the evolved programs of the route disruption attack have a perfect detection rate and false positive rate versus the ad hoc flooding attack, as shown in Table 5.9. Hence, it is proven by using the proposed algorithm.

5.18 Conclusions

A mobile ad hoc network is a collection of mobile nodes with a self-configured network. These networks have no fixed and preestablished infrastructure without any centralized administration or base stations.

Providing security in mobile ad hoc networks is of prime concern due to the need of providing protected communication between mobile nodes in a hostile environment.

An intrusion is defined as any set of actions that attempt to compromise the integrity, confidentiality or availability of a resource. An intrusion detection system is a program that analyzes what happens or has happened during an execution and tries to find evidence that the computer has been misused. The use of mobile ad hoc networks has increased manifold. The security in MANETs has become of paramount importance. Ad hoc networks are vulnerable to many attacks because of their fundamental characteristics, such as lack of centralized control, dynamic topology, limited resources, and open media.

These features present new challenges for intrusion detection techniques and achieving security in ad hoc networks becomes more difficult. This flexibility introduces new security risks. Intrusion detection is an indispensable part of any security system.

The soft computing and artificial intelligence techniques presented here are used more significantly and efficiently for the evolution of intrusion detection systems. The GE technique inspired by natural evolution is explored to detect attacks on MANETs, such as ad hoc flooding and route disruption attacks. The programs are evolved using grammatical evolution written in BNF grammar. The evolved programs show good performance on simulated networks with varying mobility and traffic patterns. The mobility-related and packet-related features are used as inputs to the evolution system. The proposed GE technique and algorithm is efficient in detecting ad hoc flooding and route disruption attacks.

- The IDS alerts efficiently detect the ad hoc flooding and route disruption attacks even with a low false positive rate.
- The classification accuracy of ad hoc flooding and route disruption attacks is found the best train achieved after 1000 generations. This trend continuously maintained up to 2000 generations until the maximum (high) fitness is achieved.

This is the evidence that the soft computing technique significantly eliminates and protects intrusions by using grammatical evolution IDS programs even though complex environments.

Acknowledgments

The authors would like to thanks Professor A. Govardhan, Department of Computer Science and Engineering, JNTU Hyderabad, and Professor SSVN Sarma, Vaagdevi College of Engineering, Warangal for providing the pseudocode and feature selections.

Funding

This work is supported by the Department of CSE–Vaageswari College of Engineering, Karimnagar, Telangana-India, under Departmental grant VGSEK/S4-01/18.

References

1. Dr. L.A. Zadeh. Some reflections on soft computing, granular computing and their roles in the conception, design and utilization of information intelligent systems. *Soft Computing*, April 1998, Vol. 2, Issue 1, 23–25. https://link.springer.com/article/10.1007%2Fs005000050030.
2. A.E. Eiben, J.E. Smith. *Introduction to Evolutionary Computing*. Springer, 2003.
3. W. Banzhaf. *Genotype-Phenotype Mapping and Neutral Variations: A Case Study in Genetic Programming*. Proceedings of the International Conference on Evolutionary Computation: The Third Conference on Parallel Problem Solving from Nature, LNCS 866, 322–332, Springer, 1994. http://web.cs.mun.ca/~banzhaf/papers/GPTP3.pdf
4. Michalewicz, Z. *Genetic Algorithms, Evolution Programs*. Springer Verlag, Berlin, 231–236, 1996.
5. D. Danial Denning. An Intrusion-Detection Model. *IEEE-Transactions on Software Engineering* Vol. SE -13, No. 2, 222–232, 1987.
6. C. Perkins, E. Belding-Royer, S. Dan, Ad Hoc On-demand Distance Vector (AODV) Routing. RFC 3561, July 2003.
7. Chowdhuri, S., Roy, P., Goswami, S., Azar, A.T., Dey, N. Rough set based ad hoc network: A review. *International Journal of Service Science, Management, Engineering, and Technology (IJSSMET)*, Vol. 5, No. 4, 66–76, 2014.
8. Chowdhuri, S., Chakraborty, S., Dey, N., Azar, A.T., Salem, M.A.M.M., Chaudhury, S.S., Banerjee, P. Recent research on multi input multi output (MIMO) based mobile ad hoc network: A review. *International Journal of Service Science, Management, Engineering, and Technology (IJSSMET)*, Vol. 5, No. 3, 54–65, 2014. http://www.bu.edu.eg/portal/uploads/Computers%20and%20Informatics/Scientific%20Computing/4983/publications/Ahmad%20Taher%20Azar_Recent-Research-on-Multi-Input-Multi-Output-(MIMO)-based-Mobile-ad-hoc-Network--A-Review.pdf
9. Charles E. Perkins, Pravin Bhagwat. *Highly dynamic Destination-Sequenced Distance-Vector Routing (DSDV) for Mobile Computers*. 1994 ACM 0-89791-682-4/94/0008.
10. Mukherjee, A., Dey, N., Kausar, N., Ashour, A.S., Taiar, R., Hassanien, A.E. A disaster management specific mobility model for flying Ad-hoc network. *International Journal of Rough Sets and Data Analysis (IJRSDA)*, Vol. 3, No. 3, 72–103, 2016. https://www.igi-global.com/article/a-disaster-management-specific-mobility-model-for-flying-ad-hoc-network/156480.
11. V. Bapuji, R.N. Kumar, A. Govardhan, S.S.V.N. Sarma. Soft Computing and Artificial Intelligence Techniques for Intrusion Detection System. *Networks and Complex Systems*, ISSN 2224-610X (Paper) ISSN 2225-0603 (Online), 24–31, Vol. 2, No. 4, http://www.iiste.org/Journals/index.php/NCS/article/view/2374/2373.
12. C. Ryan, J.J. Colline, M. O'Neill. Grammatical evolution: Evolving programs for an Arbitrary Language. *Proceedings of the 1st European Workshop on Genetic Programming* LNCS 1391, 83–95, Springer, 1998.
13. M. O' Neill, C. Ryan. *"Grammatical Evolution: Evolutionary Automatic Programming in an Arbitrary Language"*, Springer, 2003.
14. Ns-2: The Network Simulator. http://www.isi.edu/nsnam/ns.
15. Wroblewski. Finding Minimal Reducts Using Genetic Algorithms. *Proceedings of the Second Annual Joint Conference on Information Sciences* 186–189, 1995.
16. Yang J., Honavar V. Feature Subset Selection Using Genetic Algorithm. *IEEE Intell Systems* 12(2), 44–49, 1998.

6

Introduction to Coverage Optimization in Wireless Sensor Networks

Huynh Thi Thanh Binh and Nguyen Hai Nam

CONTENTS

6.1 Wireless Sensor Network Architecture

Recently, the Internet of Things (IoT) has received huge interest from scientists and technology designers throughout the world (Feng Xia et al., 2012) and is considered the next possible focus of network generation in the near future (Luigi et al., 2010). This concept starts from the idea of building a network consisting of all real-life objects (things) in which they can connect, interact, incorporate, communicate data and information, accomplishing complex tasks (Luigi et al., 2010; Khalil et al., 2014). This way, the strength of the Internet is fully investigated in a highly effective way, bringing in practical applications in many fields such as economy, information technology, military, etc. In practice, some specific applications of the IoT, namely smart homes, smart traffic (Figure 6.4) and so on have proved the feasibility, convenience and promising benefits that the technology might bring about.

However, the IoT can only be successfully applied with the support of sensor networks (SNs). This is understandable, since every object used in an IoT system has to manage a specialized sensor that monitors data and transfers sensing information during its tasks. A SN can be modelled by a net consisting of multiple sensor nodes, each of them is equipped with one or more sensors that detect and track changes of physical variables such as light intensity, temperature, air pressure, saltiness, noises and so on in a predetermined area called the monitoring area. Each sensor node consumes limited electricity sources, which are usually batteries. Data, after being collected by sensor nodes, will be converted into digital signal and be either exchanged with other nodes or transmitted to the base station (or sink node). The general architecture of a WSN system is illustrated in Figure 6.1. In this way, SNs bring a revolution to the means of collecting and transmitting environmental data due to great improvement in reliability, signal strength and other factors in the network. In comparison with wired sensor networks, wireless Sensor Networks (WSNs) have many advantages. WSNs can be deployed more easily and thus, they are more economical in term of deployment costs. For that reason, WSNs are now the major cutting-edge technology applied in a practical IoT system (Khalil et al., 2014).

Today, the development of the Internet, radio communications and information technology has brought potential development of sensor networks. The sensor network might include different types of sensors, such as earthquake sensors, magnetic field sensors, low-rate sampling sensors, camera sensors, infrared sensors, sound sensors, radar and so on, which can monitor large areas under different conditions. We will take a look at some of WSNs' applications in order to understand their necessity.

- Application in security and military nature: In response to diseases and natural disasters, a large number of sensor nodes are released from above, which will tell us the exact location of danger zones. Due to the high-density, randomly deployed WSNs, the observer is able to have the correct information and ensure efficiency searches with the investigation of soft computing and machine learning. The use of sensor networks also helps people work in harsh or even deathly dangerous places.

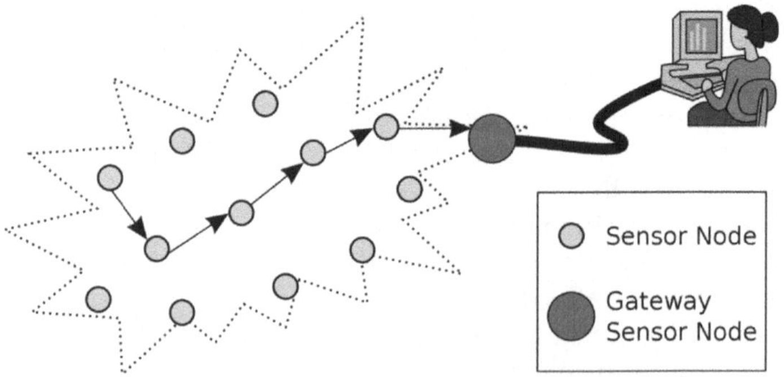

FIGURE 6.1
A typical WSN architecture.

- In security applications as intrusion detection and the arrest of criminals: Sensor networks in the military detect and contain information about the movements of enemies, explosives and other information. WSN systems are also used for detecting and classifying chemical and biological substances, radio waves, and nuclear radiation.
- Disaster Forecast: Sensor networks have been widely studied and applied in the field of disaster forecasting and proved to be a promising research direction, helping to predict disaster effectively when combined with soft computing and machine learning investigation. Drought forecasting system was proposed in Hsu-yang et al. (2006), which introduced a model based on a Drought Forecast and Alert System (DFAS), using the data of rainfall volume, average temperature and soil moisture in previous days to predict the extent of drought in the next seven days by applying neural network technology.
- Applications in Smart Home: Connect and control smart in-home devices as well as security monitoring.
- Vehicle and Related Information Monitoring: The objective of this system is to collect information through the sensors, processing and storing it in its centre and use it for necessary cases. The system was installed along the main roads and collected traffic-related information such as traffic density, traffic congestion, number of vehicles, traffic flows. The system will monitor continuously and provide real-time information, as shown in Figure 6.2. The information collected is used for coordinating traffic flows and other purposes.

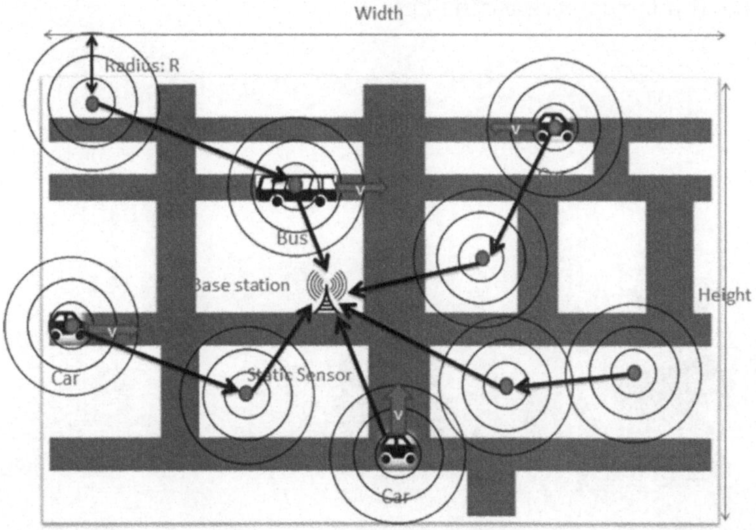

FIGURE 6.2
Traffic monitoring.

- Applications for Automatic Building: Managing multiple systems at the same time such as managing staff, electricity, heat, light and so on by attaching the chip to the devices from which can be controlled simply by mobile phone or the central computer.
- Applications in the Industrial Automation Management Process: Management, control, efficiency and safety in industrial activities; the sensors are placed in the working area to monitor the working process and detect if any incident happens. The collected data will be transmitted to the central station so as the management can measure and take action in time.
- Applications in Medicine: Diagnostics, health care and treatment of diseases are typical applications of WSNs in medicine; attaching devices directly to the patient to measure regularly the information on blood pressure, heart rate and so on are also promising directions for medical treatment, which helps hospitals to shorten treatment time and provide more accurate diagnosis. In doing so, doctors can treat patients in an effective way.
- Applications for Smart Grids (Figure 6.3): Currently in Asia, a Chinese company named Honeywell has deployed a smart grid system in Tianjin, which includes equipment for load balancing (demand response), which is deployed in many households as well as in business, industrial and administrative areas. The legislation on the smart grid has also been issued in Australia, New Zealand, Singapore, South Korea and Japan. In other Asian countries, including Vietnam and Thailand, the smart grid system only appears in the fundamental research phase.

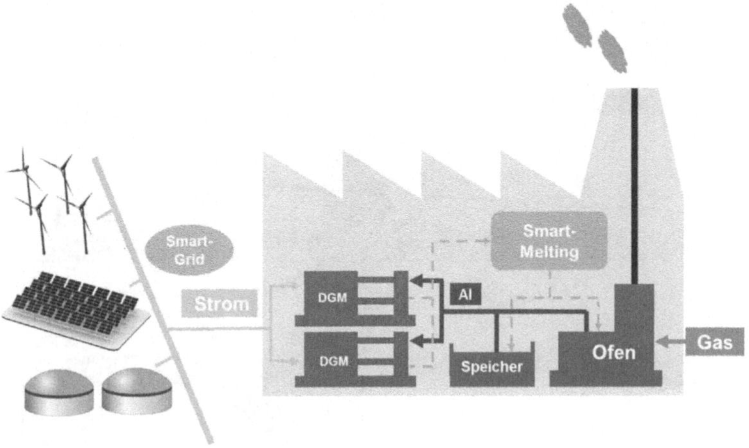

FIGURE 6.3
Smart grid applied in smart melting.

- Therefore, international collaboration plays a very important role in the development of smart grids in Asian countries, promoting research on sensor networks and its application in smart grids.
- Applications for Smart Agriculture: Installed sensor networks that are scattered in the farming fields continually gather information to check when agricultural products reached harvestable maturity. In Egypt, such a system has been implemented, using the APTEEN protocol which is current best protocol for accurate harvest, bringing remarkable results (Sherine and Basma 2013). Investment costs for potatoes accurate harvest solution is applying WSN is payback within one year, the network prolongs for 6.5 months of life expectancy, which is longer than the usual Egyptian potato crop season (6 months). The crop period to harvest potatoes is shortened from 6 months to 3 months, doubling the current economic benefit. Besides Egypt, China has also applied WSNs to improve the efficiency of greenhouse cultivation, improving the safety and quality of agricultural products (Sheng et al., 2015).

The more frequently WSNs are applied in real-world systems, the higher the expectations that users and network designers have. WSN designing nowadays needs to satisfy a wide range of standardized evaluation criteria in which the vast development of information technology also plays a significant role. In other words, WSN deployment depends on several factors, as follows:

- Network Tolerance: Some sensor nodes can be deactivated due to power shortages, physical damage or environmental impacts. The tolerance is defined as the network's ability to bear against the dead nodes' presence, continuing its normal routine and maintaining the custom-built functions.
- Scalability: In practice, WSNs may be deployed with hundreds or thousands of sensor nodes or even more, depending on the area of the monitoring field and the accuracy and reliability requirements of each application. Because of that, it is required to have a scalable network structure so that the system can be extended to the desired scale, corresponding to specific applications.
- Equipment Cost: Since the network may contain a great number of sensors, the cost of one sensor node may strongly affect that of the entire system. Keeping the cost for one individual sensor node could help reduce the equipment cost.
- Hardware Integration: Each sensor node is constrained to the following hardware requirements: small size, low energy consumption, low manufacturing cost, environmental adaptability, self-configure ability and automatic operation.

- Adaptability to Different Operation Environments: Even though sensor nodes are sensitive and vulnerable to environmental changes, they might be placed with high density in polluted, poisoned or underwater environments. In such circumstances, it is crucial for sensor nodes to adapt to multiple types of environments.
- Proper Transmission Method: Wireless transmission mediums between sensor nodes may be radio waves, infrared or optical media. To establish a common unified operation for the network, it is necessary to have a global standard for a wireless transmission medium.
- Configure Network Sensor: Sensor networks consist of a large number of sensor nodes; thus, there appears a need to establish a stable configuration.
- Energy Consumption: Each sensor node is equipped with a limited power source. In some applications, the addition of energy is impossible. Therefore, the network lifetime depends on the lifetime of the sensor node, which is dependent on the lifetime of the battery. For that reason, scientists are now working to figure out the proper designing algorithm and protocol for sensor nodes to save their limited power source.

Despite a variety of practical applications in many different fields in real life, the implementation and deployment of WSNs are facing several difficulties relating to both the sensors' physical properties and the expectation of a well-operating network system. Additionally, the requirements for a WSN system increase in a technical compared to what they used to be. In particular, for WSN applications to be widely used in multiple fields, there are many challenges that must be overcome:

- Data Storage: As the sensors sample data from the environment continuously, this data is not stored permanently because of limited storage capacity of the sensors. To save up storage, data is compressed, filtered, and aggregated from the input, and the old data must be erased. Data has to be forwarded to the central server before being saved in the network.
- Energy Consumption: Energy conservation is an important problem for sensor networks. In large-scale networks, energy monitoring and supplying power to the entire network is not possible. Therefore, the algorithms and techniques to minimize power consumption should be investigated.
- Network Tolerance: Some sensor nodes can be deactivated due to power shortages, physical damage or environmental impacts. The tolerance is defined as the network's ability to bear against dead nodes' presence, continuing its normal routine and maintaining

the custom-built functions. Therefore, the design of WSNs with high fault tolerance is also a difficult problem for developers and researchers, such that the system operates normally under harsh environmental conditions and natural disasters.

- Positioning: Using WSNs to locate or track events are a matter of concern at present. Integrating GPS in the sensor nodes results in the absolute position of the node but it produces great wastes in both energy consumption and equipment cost, and thus is inefficient. Meanwhile, other positioning techniques come up with a relative node position, which is less reliable than that of GPS. Therefore, the problem of how to determine the location of the sensors remains as an important and challenging issue.
- Scalability: In practice, WSNs may be deployed with hundreds or thousands of sensor nodes or even more, depending on the area of monitoring field and the accuracy and reliability requirements of each application. Because of that, it is required to have scalable network structure such that the system can be extended to the desired scale corresponding to specific applications.
- Security: The temperature information collected by environmental monitoring applications seems to be harmless, but is not. The operation of a building can be observed easily by analysing information about the temperature and lighting of the building. This piece of information can be used to organize an attack plan on a company. Therefore, WSNs should be able to keep sensing information secretly. In security applications, data security becomes very important not only for maintaining confidentiality, but because monitoring data must also be able to authenticate the data transmission channel. The combination of secrecy and authentication are essential requirements of all three types of applications. The use of encryption and decryption will increase the cost of energy and bandwidth. Data encryption and decryption must be transmitted with each packet, which affects the performance by increasing the packet size and energy consumption, as well as shortening lifetime expectations.

Besides the previously mentioned defects, there exist some typical weaknesses of sensor nodes, such as limited power sources, physical vulnerabilities and restricted monitoring areas, leading to the difficulties in WSN deployment in term of managing a network lifetime, coverage and so on. There are many studies, both in Vietnam and worldwide, that have investigated those disadvantages of WSNs in order to indirectly improve the network lifetime. While the majority of research focuses on energy optimization, the others study coverage optimization, both of which support the underlying aim of network lifetime improvement. According to a survey conducted by

Bang Wang et al. (2011), coverage is the most important factor in WSN quality evaluation since it reflects how well an area is monitored. For that reason, there have been many studies following this approach.

6.2 Coverage Optimization in WSNs

A survey conducted by Anju and Rishi (2015) describes multiple concepts in the coverage problems of WSNs and related works. It showed that the sensing coverage strongly affected the quality of service in WSNs. The motivation of this problem is to monitor every single piece in the deploying region by at least one sensor node. It showed that an effective coverage deployment strategy prolongs the network lifetime when sensor nodes have a very limited energy source. According to the survey, many strategies based on computational geometry were investigated and proposed to solve WSNs coverage problems.

From the above analysis, it is clear that studying and proposing effective optimization strategies for solving coverage problems in WSNs has received attention worldwide. There are a number of research groups studying this problem. However, each application in wireless sensor networks has a different objective in term of coverage. The coverage problems in WSNs can be modelled differently with respect to various purposes and applications, but all of them fell into three categories: area coverage, target coverage and barrier coverage; each of them corresponding to one or more specific applications of WSNs in practice. For each subcategory, the problem modelling features distinctive properties of the network, and the network implementation and deployment also require different evaluation criteria. For that reason, it is necessary to study each of three coverage problems in depth in one WSN.

6.2.1 Area Coverage

It is not until 2013 that Yourim Yoon and Yong-Huyk provided a new model for the area coverage problem: for a predefined number of sensor nodes with different sensing ranges, one must find their optimal locations so that they cover up the largest area possible. This model is illustrated in Figure 6.4, in which, the term 'coverage' refers to the sensing coverage and is defined by a perfect circle disk centred at the sensor node having its radius equal to the sensing range. This is an NP-hard combinatorial optimization problem. The authors showed that it is likely unable to solve this problem using deterministic algorithms such as circle packing (Andrew Howard et al., 2002) since sensor nodes have different sensing radii. Thus, Yourim Yoon and Yong-Huyk (2013) proposed a genetic algorithm for finding a proper deployment topology to tackle the issue. Specifically, an individual is an array of sensor

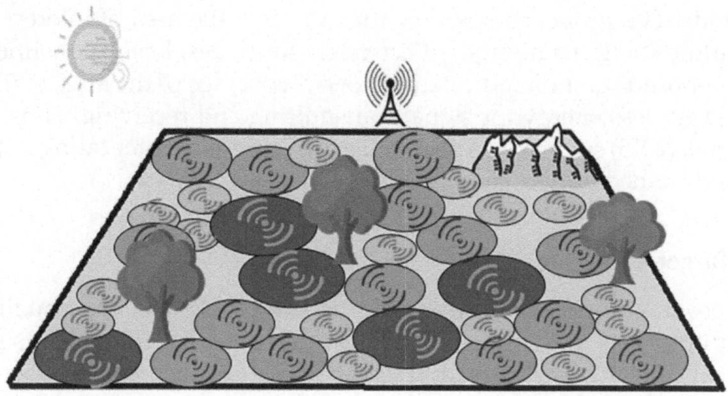

FIGURE 6.4
Area coverage.

nodes' coordinate, each member of the array shows the actual position of the corresponding sensor node in reality. The sensor nodes are in descending order by sensing range. An individual X will be evaluated by its coverage area (CoA) corresponding to its real coordinate, which is computed by the Monte Carlo technique. The idea behind the Monte Carlo method is to randomly create a very large number of points (L points) and check if those points are in sensor node X's coverage area. The complexity of this fitness method is $O(nL)$ with $L >> n$. The experimental results show a great improvement in the quality of area coverage yet they did not reach the optimal arrangement. Thus, it is possible to further optimize the algorithm to achieve better results.

In early 2015, Dinh et al. proposed a new fitness function that has the complexity of $O(n^2)$ in order to improve the calculation time of the algorithm presented in Yourim Yoon and Yong-Huyk (2013). In addition, a heuristic initialization is applied to create a better first generation of the genetic algorithm (GA), enhancing the quality of the final results. Moreover, a modified version of the original mutation is investigated in the algorithm, which utilizes the dynamic Gaussian function for replacing the static one in Yourim Yoon and Yong-Huyk (2013). Thanks to those mentioned modifications, the new genetic algorithm not only derived a better set of experimental results but also has the computational complexity reduced greatly from $O(nL)$ with $L >> n$ to $O(n^2)$. Since the dataset for testing (Dinh et al., 2015) consists of a small number of sensor nodes and most of them are static nodes (nodes that cannot move), we expect to further ameliorate the algorithm so as to deal with more sensor nodes including both static and moving ones.

In another approach, Sudip Misra et al. (2015) addressed the problem of network coverage and connectivity and proposed an efficient solution to maintain coverage, while preserving the connectivity of the network by keeping each sensor node stay in the communication range of at least one other node. The proposed solution aims to cover the area of interest (AOI), while minimizing the number of active sensor nodes. Results of simulation of the proposed solution indicated up to 95% coverage of the area, while consuming very less energy for signal transmitting and receiving. Thus, Sudip Misra et al. (2015) suggested a multi-objective approach integrating both coverage and connectivity to improve the quality of the WSNs.

6.2.2 Target Coverage

Target coverage optimization problem is the problem of maximizing the number of targets in the monitoring field that are covered, which is shown in Figure 6.5. Many strategies for sensor network deployment have been proposed to optimize the target coverage in WSNs, such as applying the virtual force algorithm (VFA) to find optimal positions for each and every sensor node or utilizing computational geometry techniques, including mesh establishment and Voronoi geometry. However, using metaheuristic techniques and soft computing algorithms have shown their potential in solving the coverage optimization problem best amongst other methods.

In 2010, Salma Begum et al. used the artificial bee colony (ABC) algorithm to find the smallest set of sensor nodes that cover all targets in a given area. The experiments are performed under complicated circumstances when some targets need to be monitored with high reliability by more than one sensor (the k-coverage problem). In general, it is shown that swarm algorithms like ABC and PSO (particle swarm optimization) can be applied effectively for both area coverage and target coverage problems in 2D and 3D terrains.

In 2013, Omar Banimelhem et al. presented a sensor network consisting of both static and mobile nodes. In the beginning, a network is initialized by

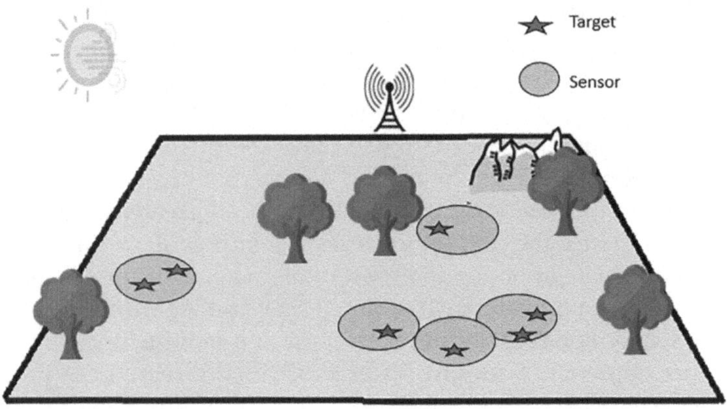

FIGURE 6.5
Target coverage.

randomly placing all static sensor nodes into the monitoring area. After that, there appear holes and vacancies in the region, at which none of the targets can be monitored by the deployed static sensors. Thus, mobile sensor nodes are added to the field to cover up the blank spaces. The research in Omar Banimelhem et al. (2010) proposed a genetic algorithm (GA) to search for the smallest number of mobile sensors and their best locations such that the target coverage is maximized. In 2015, Ta et al. proposed an improved genetic algorithm to tackle the target coverage problem and the experimental results show a significant improvement compared to the one presented in Omar Banimelhem et al. (2010).

6.2.3 Barrier Coverage

In 1992, Gage defined the concept of barrier coverage for the first time in robotic sensors. The term barrier refers to a line or a curve going across a monitoring field such that nothing shall be passed by it without being detected. Figure 6.6 illustrates such a system. In 2005, Kumar et al. provided a definition for the k-barrier coverage problem, and it proposed an optimization algorithm to determine if a field is k-barrier or not: An area is considered to have k-barrier coverage when any intruder attempting to break into has to go by at least k sensors. In 2008, Liu et al. proposed an effective distribution algorithm to generate a strong barrier randomly deployed on a geometric strip (a rectangle with long width) area. In 2009, Saipulla et al. studied about barrier coverage in line-based deployed WSNs and established a tighter lower bound for the existence of barrier coverage compared to former studies. In 2011, the problem of finding the smallest number of camera sensors in a camera SN that can form a barrier is studied by Wang and Cao (2010). In 2012, Ma et al. proposed a new method for selecting some nodes from a random distribution camera sensor network to form a camera barrier, and Tao et. al. (2012) examined finding proper sensor directions to establish a strong barrier. In 2014, Wang studied the one-barrier problem, the establishment of a barrier, and the smallest number of mobile sensor to form one barrier.

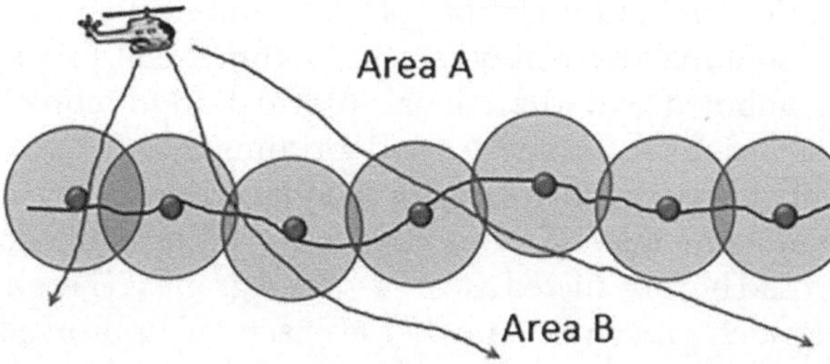

FIGURE 6.6
Barrier coverage.

That research also involved the efficiency of barriers in hybrid sensor networks with several static nodes initialized at the beginning, the minimal cost for moving mobile sensors and so on.

From the previous discussion, the most challenges in WSN systems are formulated as combinatorial optimization problems and are classified as NP-hard problems (Bang Wang, 2011; Anju and Rishi, 2015). There exist two approaches to solve NP-hard problems: exact solutions and approximate solutions. The first approach includes branch and bound, branch and cut, branch and price, dynamic programming, linear and integer programming, Lagrange relaxation and so on. Meanwhile, the second approach consists of simulated annealing, tabu search, iterated local search, variable neighborhood search, genetic algorithm, memetic algorithm, etc.

On the one hand, exact algorithms are guaranteed to result in exact solutions with a calculatedly limited computational time, and the solution is proven to be optimal with respect to experimental data sets at a specific range. In the case of no solution available, exact algorithms should be aware of and prove the nonexistence of a valid solution. By applying these algorithms, it is always possible to get the concave or convex of the optimal solution, even when the algorithm stops before its end. However, in most practical cases, the computational time of exact algorithms vastly increases when the data size augments (usually exponentially). Thus, this approach is not feasible to solve large-scale problems or scalable problems due to limited physical memory storage. In other words, it is difficult to upgrade the algorithm regarding the parameters changing or problems scaling up.

On the other hand, approximate approaches using soft computing returns nearly optimal solutions instead of the optimal one. However, these algorithms usually provide an acceptable computational time for all possible experimental datasets. Currently, there is no technique to solve NP-hard combinatorial optimization problems with big data size that offers the exact solution in acceptable time duration. Soft computing is one approach to solve popular NP-hard problems that receives great interest recently. Many studies show that soft computing offers a potential paradigm for solving global optimization problems (Luigi et al., 2010; Gage, 1992; Kumar et al., 2005; Ma et al., 2012; Tao et al., 2012). However, it is demanding to do in-depth studies about the effectiveness and feasibility of integrating different techniques to provide better solutions for combinatorial optimization problems (such as local search) combined with directional information to refine findings and combination of various soft computing algorithms.

In conclusion, it is possible to classify mentioned problems in distributed and software systems into NP-hard combinatorial optimization problems; most of them could be formulated as large-scaled graphs. There are two methods for solving such problems, namely the exact and approximate ones, but we are in favour of the latter approach for two reasons. First, metaheuristic algorithms offer acceptable time complexity when tackling expensive optimization problems. Second, there are many functional and non-functional requirements (e.g. low power consumption, maximized coverage, etc.) in distributed and software systems, such that particular heuristic algorithms designated for each requirement are on-demand. To be specific, we will mainly focus on studying various metaheuristic algorithms (e.g. EA), which help handle both single objective and multiple objectives optimization problems in distributed and software systems.

6.3 Literature Review

Coverage is viewed as a standout amongst the most critical elements contributing on the evaluation of the quality of WSNs. Accordingly, the problem of coverage optimization has been widely studied, and therefore, numerous exploration works have been directed on this matter (Bang Wang, 2011).

According to a survey, there are six major designing factors contributing to the coverage measure of WSNs systems. They are coverage type, activity scheduling, coverage degree, coverage radio, network connectivity and deployment method (Bang Wang, 2011). Based on that, it is possible to classify the coverage problems in WSNs into three subclasses, namely area coverage, target coverage (also known as point coverage) and barrier coverage.

Area coverage optimization problem was first approached by using a potential field-based method called VFA, and the goal was maximizing the total area coverage of all sensor nodes (Dinh et al., 2015). The basic step of this algorithm is that sensor nodes push each other away, spreading the nodes evenly throughout the field. This step is repeated until no further improvement of the total coverage of nodes is recorded. The novelty of this method is that it could adapt to monitoring fields of arbitrary shapes. However, this model remains useful only in theory since many impractical assumptions were made such as all the sensor nodes are of the same type and the only reason for downtime is due to out of battery (in reality, sensor nodes could be physically damaged).

Discussing coverage optimization with more in-depth elaboration, Anju and Rishi (2015) came up with a more realistic model for this problem, which is to cover the entire given field with an arrangement of small-density sensor nodes. Two placement strategies called deterministic node placement and random node deployment were introduced to solve the problem with support of mathematical analysis and arguments.

Later, another model for the area coverage problem was proposed to search for the optimal placement locations for a predefined number of sensor nodes that yields the largest sensing coverage (Yourim and Yong-Huyk, 2013). In this model, a more realistic constraint is added such that there are different sensing ranges in the network, which stems from the fact that WSN systems are usually extended with new and more powerful models of sensors which possess longer and stronger sensing/transmitting ranges. The term coverage in this model denotes the sensing coverage and that of a sensor *s* is defined by a perfect circle centred at the *s*, whose radius equals its sensing range. A genetic algorithm is proposed to find the optimal deployment topology to maximize the coverage, since the problem is NP-hard and is more complex with a more realistic constraint added. Going into detail, this work defines an individual as an array of sensor node coordinates, each element of which designates the actual position of the corresponding sensor node. Sensor nodes are in descending order by sensing range. The evaluation of one individual is done by its coverage area (denoted *CoA*) which corresponds to its coordinate in reality and is computed by the Monte Carlo technique. As for calculating the coverage of a sensor node *s* with a sensing range *r* and a point *p*, it is possible to tell whether *p* is within the *s*'s coverage, represented by a Boolean value *b*. The idea behind the Monte Carlo method is to repeat this step many times to approximate the coverage area of *s* by randomly creating *L* points (*L* is a very large value) and accumulate the Boolean value *b*. This method for computing the fitness of each individual in the genetic algorithm has a great complexity of $O(nL)$ with $L >> n$, where *n* is the number of sensors (Yourim Yoon and You-Huyk, 2013). With this approach, it is possible to cover up to 90% of the upper bound of the possible area coverage calculated from the sensing range of sensor nodes, the number of nodes and the given field's area. However, there remains a great difference between the achieved coverage and the optimal solution, illustrated by about 10% offset from the upper bound. Another weakness of the solution is its slow runtime. Another work following up this one reported that the runtime could take up to nearly an hour for inputs of large size. Both weaknesses stem from the expensive evaluation function when $L >> n$. This could be lowered in complexity by choosing a smaller value of *L*, but this leads to a less accuracy result. Since there are many studies attempting to improve this result, the best algorithm in this study is OTPHGA, which is OTPGA (genetic algorithm with normalization) hybridized with VFA (virtual force algorithm) (Yourim Yoon and You-Huyk, 2013). OTPHGA will be regarded as the state-of-the-art algorithm and recent solutions offering better performance will be presented in Section 6.4.

It is shown that the problem formulation for the coverage optimization problem is well addressed and OTPHGA algorithm has been effectively used to tackle it. However, there exist weaknesses with both the problem formulation and the performance of the solution. The solution delivered by OTPHGA has yet reached the upper bounds with regards to big instances and requires a long execution time (up to almost an hour for large instances). Therefore, it is possible to improve the model with more realistic constraints and develop better methods and performance in terms of both quality of solution and runtime. Many improved methods are proposed and compared to OTPHGA, which makes it the state-of-the-art algorithm for this problem formulation of area coverage optimization.

As for a target coverage subclass, one of the most popular problems is to find the minimum number of sensors so that all the points/targets in a given field are covered. Various methods, including genetic algorithm and greedy algorithm, were proposed to tackle the problem (Bang Wang, 2011). Additionally, problem formulations where network lifetime maximization is considered along with coverage optimization were also introduced, such that the coverage is guaranteed with connectivity constraints satisfied. The MCT problem (maximum cover tree) is one well-known branch for the connected target coverage optimization problem, in which, a tree to represent a subset of active sensors is utilized. This NP-complete problem was solved using some heuristic algorithms by configuring an alternative path to transmit data in case a node is down, which increases the system's tolerance to faulty incidents (Zhao et al., 2006 and 2008).

Another approach to the connected target coverage problem is to enforce *k*1-coverage and *k*2-connectivity requirements in a way that *k*1 different active sensors cover each of the targets, and a *k*2-connected network could be formed from the set of active sensors. A heuristic algorithm to solve this problem was introduced (Li et al., 2007); however, the number of both the targets and sensors used are very small given the sensors' position, which suggests its impracticality. When working with larger with more dense distribution of targets, this algorithm is not sufficient to tackle the problem and a more efficient one is needed.

Besides coverage maximization, the problems of coverage maximizing and network lifetime lengthening combined with connectivity constraints were widely studied as well; 'connectivity' meaning the ability to communicate between sensor nodes. Two sensors s_1 and s_2 can transmit and receive data from/to each other if the distance between s_1 and s_2 is less than both of their radiuses. A connectivity-guaranteed target coverage model where two sensors can communicate when the distance between their locations is less than or equal to the suggested sensor radius (Kar and Banerjee, 2003). This model received interest from the research community since it could be approached as either a constrained optimization problem or a multi-objective optimization problem (Bang Wang, 2011). Besides, the target coverage with connectivity fault-tolerance problem has been investigated, which provides some needed background knowledge (Li et al., 2007). However, small input size with not many sensor nodes and targets plus the predefined node placement reduce the ability to generalize the results obtain from the experiments to that in real life.

In conclusion, it is shown that the models proposed for the coverage optimization problem is not very practical, since the attempt to add more realistic constraints to the model has been done in various studies. For future research, a plan to examine obstacles with a variety of barriers in WSNs as well as to work on the connectivity constraints in order to fully reflect practical scenarios of area coverage optimization problems in WSNs have the most potential. Better heuristic/soft computing techniques should also be employed to compensate the additional computation resulting from new realistic constraints.

6.4 Problem Formulation

6.4.1 Model 1 – Conventional Coverage Optimization

To start with, we define the given field as a 2D domain A sized $W \times H$. Given a number of sensors, the goal is to cover the field as much as possible. There are k types of sensor nodes. A type i sensor can sense events within a radius r_i (referred to as sensing range). This is referred to as a binary sensing model and is illustrated in Figure 6.7. The goal here is to find the best placement scheme for all sensor nodes, in which, 'best' is defined as the highest value of area of coverage on A (denoted coA).

The problem of maximizing the area coverage in WSNs is formulated in detail below.

Input:

- W, H: the width and the length of the given 2D domain A.
- k: the number of sensor types
- n: the total number of sensors
- n_i: the number of sensors for type i ($i = 1, 2, \ldots, k$), such that:

$$\sum_{i=1}^{k} n_i = n \tag{6.1}$$

- r_i: the sensing radius of sensor type i ($i = 1, 2, \ldots, k$)

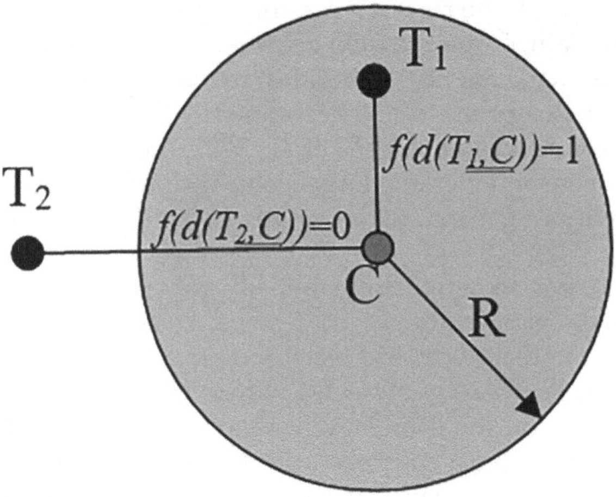

FIGURE 6.7
Binary sensing model.

Output: The position of the sensor node set
Objective: Maximizing *coA*, whose value is calculated as:

$$coA = area\left(\bigcup_{i=1}^{k}\bigcup_{j=1}^{n_i} c_{r_i}\left(x_{i_j}, y_{i_j}\right) \cap A\right) \rightarrow max \tag{6.2}$$

with $c_{r_i}\left(x_{i_j}, y_{i_j}\right)$ is the circle at $\left(x_{i_j}, y_{i_j}\right)$ and the radius is r_i; the *area*(X) is the area of the domain X.

This model of the coverage maximization problem was shown to be NP-hard [3] and it has been extended to different variations under other categories of optimization problems in WSNs.

6.4.2 Model 2 – Coverage with Connectivity Fault-Tolerance Model

This model differentiates from the first one in the additional of a constraint for connectivity fault-tolerance. Using the Boolean disk coverage model, given a target t as a point in the 2D domain and all sensor nodes share the same sensing radius R, the ability of sensor to cover target is described as follows:

$$f(s,z) = \begin{cases} 1, \text{ if } d(s,z) \le R, \\ 0, \text{ otherwise,} \end{cases} \tag{6.3}$$

where $d(s, z)$ is the Euclidean distance between sensor s and target z.

In the opposite, the communication range of all sensors is assumed to be twice as long as their sensing range, which is $2 \times R$. Data transmitting, relaying and receiving from/between/to others sensors are only possible if the Euclidean distance between them is within the range $2 \times R$. Function $g(s, s')$ defines the possibility that two sensors s and s' can interact with each other:

$$g(s,s') = \begin{cases} 1, \text{ if } d(s,s') \le 2 \times R, \\ 0, \text{ otherwise.} \end{cases} \tag{6.4}$$

Given a set of sensor nodes and a base station placed on the 2D domain A with $A = \{(x, y) : 0 \le x \le W; 0 \le y \le H\}$; $G = (V, E)$ be a graph with vertices set $V = \{B, s_1, \ldots, s_n\}$ and edge set E designates the connection link among sensor nodes and the base station; B denotes the base station and s_i $(1 \le i \le n)$ is a label for the ith sensor; there exists an edge $e \in E$ connecting a and b if and only if $g(a, b) = 1$; $a, b \in V$.

In this model, the base station can acquire information about a target t if there exists a path $P=\left(s_{i_1},e_1,s_{i_2},e_2,\ldots,s_{i_m},e_m,B\right)$ in graph G where $s_{i_j}\in V/\{B\}$ and the edge e_j connects two sensors s_{i_j} and $s_{i_{j+1}}$ satisfying $\left(s_{i_{j+1}}=B\right)$ and $f\left(s_{i_1},t\right)=1$.

To post the constraints so that there is a backup path connecting each target t to the base station, we need to define such two paths, in this case, $P=\left(s_{i_1},e_1,s_{i_2},e_2,\ldots,s_{i_m},e_m,B\right)$ and $P=\left(s_{i_{1'}},e_{1'},s_{i_{2'}},e_{2'},\ldots,s_{i_{m'}},e_{m'},B\right)$. To say that they are distinguishable sharing the same ends, $v_i\neq v'_j,\forall i:1\leq i\leq m,\forall j:1\leq j\leq p$ is needed. With these denotations, the problem can be formulated as follows:

Input:

- W, H: respectively the width and height of the 2D domain A
- n: the number of targets which we want to collect information
- $T = \{(x_i, y_i)|\ 0 \leq x_i \leq W, 0 \leq y_i \leq H, i = 1..n\}$: the 2D coordinates of n targets
- R: the sensing radius of all sensors
- $B(x_B, y_B)$: the 2D coordinates of the base station $(0 \leq x_B \leq W, 0 \leq y_B \leq H)$

Output: The total number of sensors used, n, and their coordinates

Objective: The minimum number of sensors on A such that between each target in t and the base station, where there are two distinguishable paths to send information

This target coverage model and one of the solution for it are illustrated in Figure 6.8 with targets located randomly on A, along with sensor nodes and a base station.

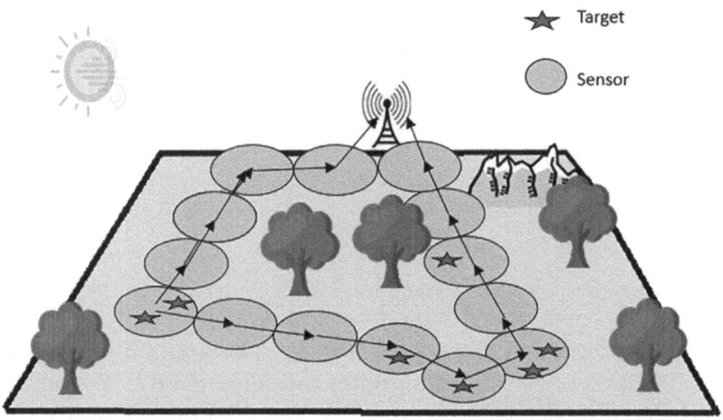

FIGURE 6.8
Target coverage model with one possible solution.

6.5 Recent Works on Coverage Optimization in WSNs

Attempting to improve the state-of-the-art algorithm OTPHGA, a new approach is taken to solve the root of the problem in the fitness function's complexity of $O(n^2)$ (Dinh et al., 2015). A new fitness function was proposed to reduce the computation power needed compared to the original version (Yourim and Yong-Huyk, 2013). The idea to reduce the complexity is simply to substitute the Monte Carlo method, which accounts for a great deal of computation. Dinh et al. introduced a new concept of the overlapping called *Olap* that computes the overlapping area solely based on the relative positioning of two sensors and their sensing range. This apparently saves the algorithm from repeating the point sampling process in Monte Carlo evaluation. Moreover, a modified mutation function utilizing the dynamic Gaussian function was also introduced. As a result, the new genetic algorithm called IGA not only produced better results but also achieved lower time complexity of $O(n^2)$ (taking more than one minute for large test data) compared to the original $O(nL)$ with $L >> n$ (Yourim and Yong-Huyk, 2013).

Later, the runtime for this problem is even further reduced to a matter of seconds by using a particle swarm optimization (PSO) and a bio-inspired swarm optimization algorithm, designated for the area coverage with obstacles. The difference between PSO and genetic algorithm is the interaction between individuals in the population (Kaveh, 2014). Each individual in PSO moves in its own fashion characterized by a velocity and an accelerator vectors, thus making use of the experiences and knowledge of the whole population to find the approximate optimal efficiently and promptly.

However, the framework of a general PSO has some weaknesses in its design so that it suffers from premature convergence (Kaveh, 2014; Nakisa, 2014). Democratic PSO (DPSO) appears to overcome this weakness, i.e., to solve the premature convergence problem and indirectly improve the quality of solution founded by PSO. Premature convergence is caused by the whole population following the best individual, leading to a possibility to get stuck in local maxima. This problem is solved in DPSO by taking bad individuals' experiences into consideration when moving the population so that all individuals are treated equally, which widens the search space significantly.

Experiments have suggested that PSO and DPSO outperform IGA and OTPHGA in all terms of quality of solution and calculation time. While PSO provides slightly better area coverage but taking a longer runtime compared to IGA's (about 50% longer); DPSO significantly outperforms the rest of the algorithms by around 60% compared to that of IGA over all instances. DPSO resulting in shorter calculation time because it converges quickly at around the 50th generation and meets the stopping condition while IGA and PSO take much longer execution time to converge.

In a different attempt to solve the coverage optimization with fault-tolerance connectivity problem, a heuristic algorithm including two phases was proposed; in which, the former seeks the smallest number of sensors to cover the target set and the latter assures the connectivity with fault-tolerance to the targets.

In phase one, the goal is to find SSCAT (sensor set covering all targets) that uses the minimum number of sensor to cover each target by at least one sensor. To simplify the problem, it is assumed that the sensors' coordinates are integers. Since this is an NP-complete set cover problem, the greedy set cover algorithm is used to solve this problem in polynomial time. After covering each target with one or more sensor(s), SSCAT set is doubled (now denoted SSCAT') to let at least two sensors covering on each target, enhancing the fault-tolerance of the target monitoring.

However, sensor nodes in this set are not yet connected. This is the responsibility of the second phase, in which sensor nodes are added to satisfy the connectivity of the network. Moreover, the connectivity fault-tolerance will also be satisfied in this phase. The fault-tolerance in WSNs is the ability to bear dead nodes in the network. In this case, sensors are distributed in a way that they form backup paths in WSNs so that when some nodes stop functioning and the network still operates as usual by utilizing the backup path (Hanh et al. 2016).

6.6 Conclusion and Discussion

Chapter 6 aimed to do a survey of most recent works in area coverage optimization problem in WSNs. It is noticeable that the models for the coverage optimization problem were improving through time with more realistic constraints added to them – from a heterogenous network to a network having multiple types of sensor nodes.

When there are multiple types of sensor in WSNs, determining the placement of sensor nodes associated with the best area coverage is the goal. IGA, PSO and DPSO algorithms were introduced to improve the performance in terms of solution quality and runtime compared to the state-of-the-art, OTPHGA. While PSO provides the best area coverage but long runtime and DPSO converges remarkably fast (although the quality of solution is not as high as PSO) IGA achieved a balance between the two evaluation matrices. Deciding which algorithm works best depends a lot on the application of the WSN system.

Following the model development, recent models are extended with the ability to suffer a small loss in the network by scheduling a backup path from every target to the base station, also known as the target coverage maximization with a connectivity fault-tolerance problem. For the future work, there are potentially countless constraints or network configurations that could be added to the model to increase its practicality. Some of them include having obstacles of specific shapes or even arbitrary shapes to the field. Besides, it is also possible to improve the algorithm to construct a network with fewer sensor nodes and a shorter running time. As for the practicality, the coverage optimization problem could be investigated with other practical constraints, such as energy and lifetime.

References

Atzori, L., Iera, A., Morabito, G. Internet of Things: A survey, *Computer Networks* vol. **54**(15), 2787–2805, 2010.

Banimelhem, O., Mowafi, M., Aljoby, W. Genetic algorithm based node deployment in hybrid wireless sensor networks. *Communications and Network* 273–279, 2013.

Begum, S., Tara, N., Sultana, S. Energy-efficient target coverage in wireless sensor networks based on modified ant colony algorithm. *International Journal of Ad hoc, Sensor & Ubiquitous Computing (IJASUC)* vol. **1**(4), December 2010.

El-Kader, S. M. A., El-Basioni, B. M. M. Precision farming in Egypt using the wireless solution sensor network technology. *Egyptian Informatics Journal* vol. **14**(3), 221–233, 2013.

Gage, D. W. Command control for many-robot systems. Nineteenth Annual AUVS Technical Symposium, 22–24, (Huntsville, Alabama, USA), 1992.

Hanh, N. T., Hanh, P. T. H., Binh, H. T. T., Nghia, N. D. Heuristic algorithm for target coverage with connectivity fault-tolerance problem in wireless sensor networks. Technologies and Applications of Artificial Intelligence (TAAI), 2016, 235–240. IEEE, 2016.

Howard, A., Mataric, M. J., Sukhatme, G. S. Mobile Sensor Network Deployment using Potential Fields: A Distributed, Scalable Solution to the Area Coverage Problem. *Distributed Autonomous Robotic Systems* vol. **5**, 299–308, 2002.

Jiang, S., Wang, W.-X., Hu, Y.-M., and Ye, Y. Design of wireless environment monitoring system for monitoring greenhouse Cultivation. *Proceedings of the 6th International Conference on Industrial Engineering and Asia Innovation Management* 219–228, 2015.

Kar, K., Banerjee, S. Node placement for connected coverage in sensor networks. International Symposium on Modeling and Optimization in Mobile, Ad Hoc and Wireless Networks (WiOpt), 2003.

Kaveh, A. *Advances in Metaheuristic Algorithms for Optimal Design of Structure*. Springer International Publishing, DOI: 10.1007/978-3-319-05549-7, 2014.

Khalil, N. et al. Wireless sensors networks for Internet of Things. Intelligent Sensors, Sensor Networks and Information Processing (ISSNIP), 2014 IEEE Ninth International Conference on. IEEE, 2014.

Kumar, S., Lai, T. H., Arora, A. Barrier Coverage with wireless sensors. *Proceedings of ACM MobiCom* 284–298, 2005.

Kung, H.-Y., Hua, J.-S., and Chen, C.-T. Drought forecast model using wireless sensor networks and frameworks. *Science and Engineering of Information Journal* vol. **22**, 751–769, 2006.

Li, D., Cao, j. Liu, M., Zheng, Y. K-connected target coverage in wireless sensor networks. The 1st International Conference on Combinatorial Optimization and Application (COCOA), also in *LNCN*, vol. **4616**, pp. 20–31, 2007.

Liu, B., Dousse, O., Wang, J., Saipulla, A. Strong barrier coverage of wireless sensor networks. *Proc. of ACM MobiHoc*, 411–420, 2008.

Ly, D. T. H., Hanh, N. T., Binh, H. T. T., Nghia, N. D. An Improved Genetic Algorithm for Maximizing Area Coverage in Wireless Sensor Networks. The Sixth International Symposium on Information and Communication Technology (SoICT 2015), 2015.

Ma, H., Yang, M., Li, D., Hong, Y., Chen, W. Minimum camera barrier coverage in wireless camera sensor networks. *Proc. of IEEE INFOCOM*, 217–225, 2012.

Misra, S., Kumar, M. P., Obaidat, M. S. Connectivity preserving localized coverage algorithm for area monitoring using wireless sensor networks. *Computer Communications* vol. **34**, 2015.

Nakisa, B., Nazri, M. Z. A., Rastgoo, M. N., Abdullah, S. A survey: Particle swarm optimization based algorithms to solve premature convergence problem. *J. Comput. Sci.* vol. **10**, 1758–1765, 2014.

Saipulla, A., Westphal, C., Liu, B., and Wang, J. Barrier coverage of line-based deployed wireless sensor networks. *Proc. of IEEE INFOCOM*, 127–135, 2009.

Sangwan, A., Singh, R. P. Survey on coverage problems in wireless sensor networks. *Wireless Personal Communications* vol. **80**(4), 1475–1500, DOI: 10.1007/s11277-014 -2094-3, 2015.

Ta, V. D., Huang, S.-C., Binh, H. T. T. Covering the Target Objects with Mobile Sensors by Using Genetic Algorithm in Wireless Sensor Networks. *Journal of Computers (JCP 2015)* vol. **10**, 300–308, ISSN 1796-203X, 2015.

Tao, D., Tang, S., Zhang, H., Mao, X., Ma, H. Strong barrier coverage in directional sensor networks. *Computer Communications*, vol. **35**(8), 895–905, 2012.

Wang, B. Coverage problems in sensor networks: A survey. *ACM Computing Surveys (CSUR)* vol. **43**(4), 32–84, 2011.

Wang, Y. and Cao, G. Barrier coverage in camera sensor networks, *Proc. of ACM MobiHoc*, 2010.

Wang, Z. Barrier coverage in wireless sensor networks. Doctoral Dissertations, University of Tennessee, Knoxville Trace: Tennessee Research and Creative Exchange, 2014.

Xia, F., Yang, L. T., Wang, L., Vinel, A. Internet of Things, *International Journal of Communication Systems*, vol. **25**(9), 1101–1102, DOI: 10.1002/dac.2417, 2012.

Yourim Yoon, Yong-Huyk Kim. An efficient genetic algorithm for maximum coverage deployment in wireless sensor networks. *Cybernetics, IEEE Transactions* vol. **43**, 1473–1783, 2013.

Zhao, Q., Gurusamy, M. Maximizing network lifetime for connected target coverage in wireless sensor networks. IEEE International Conference on Wireless and Mobile, Networking and Communications (WiMob), 94–101, 2006.

Zhao, Q., Gurusamy, M. Lifetime maximization for connected target coverage in wireless sensor networks. *IEEE/ACM Transactions on Networking* vol. **16**(6), 1378–1391, 2008.

7

Energy Efficient Cluster Formation Using the Firefly Algorithm (EECFF)

Anupkumar M. Bongale and Nirmala C. R.

CONTENTS

7.1 Introduction

WSNs consist of sensors (also called as nodes) networked through a wireless medium. The sensor nodes have a limited processing capacity and battery power. These sensors sense the surrounding information such as moisture, temperature, etc. depending upon the type of application. The sensed information is transmitted to a data collection point usually termed as a base station (BS) or sink node. In order to carry out data sensing and transmission, the sensor nodes consume their battery power. Over a period of time, the battery of sensor nodes deplete and become nonoperative [1]. There are several notable applications of WSNs in the field of gastrointestinal endoscopy [2], ECG healthcare monitoring [3], etc.

The major research challenge of WSN design is to conserve network energy and prolong the network lifetime. Presently, appreciable research is going on in this field to target better and more energy-efficient WSN design strategies. The energy utilization of the network can be controlled by proposing enhancements at the physical, link, network, transport of layers etc. of the WSN protocol stack. In Chapter 7, the design and development of cluster-based routing protocols for WSNs is proposed [4]. The design and development of energy-efficient routing protocols is one of the primary research areas concerning WSNs and there are many articles on this topic [5,6].

Routing protocols are observed to be an excellent way for prolonging the WSN lifetime. Cluster-based routing protocols especially have drawn huge attention among the research community. Recent clustering protocols for WSNs are mentioned in [7–9]. In cluster-based routing schemes, a few network nodes are elected as CHs. These nodes form a cluster with other nodes in the network. Other nodes or cluster member nodes fetch the information and send data to the respective CHs. Data is collected and aggregated at CHs and further forwarded to the BS. In general, cluster-based routing protocols operate in phases and include cluster head election, cluster formation, data transmission, data aggregation, among others. In all these phases, energy is consumed and the challenge lies in designing routing protocols that minimize energy consumption in these phases and enhance the efficiency of the network at the same time.

Typical challenges of cluster-based protocols is the election of CHs. Low energy adaptive clustering hierarchy (LEACH) [10] is a well-recognized routing scheme in the field of WSNs. LEACH has several rounds of operations. Each round in LEACH has two phases, namely the setup phase and steady state phase. In the setup phase, CH nodes are elected based on probability and the rest of the nodes (i.e., non-cluster head nodes) join the nearest available CH node to form a cluster. In the second phase (i.e., the steady state phase) CHs allocate Time Division Multiple Access (TDMA) slots to each of the cluster member nodes. The non-cluster head nodes send its sensed information in the allocated TDMA slot. The CH receives the information from its member nodes and data is aggregated at CH nodes and further transmitted to the BS. The uniqueness of the LEACH protocol is that it does not allow nodes to become CHs if they have already become CHs in one of the earlier rounds. This ensures that the burden of being a CH is rotated uniformly among all the nodes the network. Though LEACH has provided a great CH election strategy, it has its shortcomings, such as electing CHs without considering the residual energy of the nodes, the elected CHs may lie very close to one another, in some of the rounds the CHs are never elected, etc. Over the past decade, various improved versions of the LEACH protocol have been presented. LEACH protocol and its variants are covered in [9]. Apart from

LEACH, TEEN [11], PEGASIS [12], HEED [13], etc. are the few of the most popular routing techniques.

The problem of CH election in cluster-based routing protocols is that it is always considered as NP-hard. There is no definite solution for efficient CH election which suits requirements of all WSN applications. But researchers have identified that metaheuristic optimization strategies such as harmony search (HS), particle swarm optimization (PSO), fuzzy logic, artificial bee colony (ABC) optimization, genetic algorithm (GA), etc. for designing cluster-based routing protocols have optimal results. For updates on applications of metaheuristic optimization techniques in cluster-based routing protocols, readers can refer to the literature [14–16].

Metaheuristic algorithms have also gained huge popularity not only because of an effective cluster head election strategy, but also due to the provision of solutions in other WSN issues. Binh et al. in [17] have considered the maximization of the area coverage problem and proposed a solution through an improved cuckoo search and chaotic flower pollination optimization algorithms. The nodes of the network have a different sensing radius and the coverage area problem is to place the nodes at the appropriate positions of the sensing region so that the coverage area is maximized with minimum energy consumption. The authors have used nature-inspired concepts of cuckoo bird breeding and the pollination of flowers for modeling the solution for efficient area coverage in WSNs. The results obtained are compared with the genetic algorithm and PSO–based techniques based on convergence time.

Vijayalakshmi and Anandan have presented a cluster head election strategy based on the hybridization of PSO and Tabu Search algorithms [18]. The authors have identified that Tabu Search converges faster at a global solution and that PSO is applied on the obtained global solution to arrive at better search results. The hybrid approach helps in obtaining an efficient solution space and ultimately leads to the selection of efficient cluster head nodes. The optimization parameters considered for electing cluster heads are the distance between the BS and nodes and the energy level of the nodes. The presented algorithm is compared with LEAHC and PSO–based protocols, including parameters such as number of alive nodes, packet loss rate, end-to-end delay, etc.

The nature-inspired optimization algorithms are capable of providing excellent solutions if the problems to be solved are well understood and modeled properly. Apart from WSNs, there are many other domains where metaheuristic strategies have demonstrated proven solutions. The firefly algorithm (FA) designed by Xin-She Yang is one of the nature-inspired optimization techniques applied in various fields of engineering and medical science. Some of the prominent applications of FA in various domains of research are the optimization of scaling factors of ophthalmology images [19]; optimized design of a CMOS spiral inductor [20]; video frame registration [21]; optimization of proportional-integral-derivative controller parameters such as proportional gain, integral gain, derivative gain, etc. [22]; optimization of low noise amplifier parameters [23]; and many more. With such vast applications, it can be conveyed to the readers of this book that FA and other nature-inspired algorithms could be applied to any domains where optimization is a crucial aspect. In Chapter 7, a cluster-based routing protocol incorporating FA is presented. Section 7.2, provides brief information about the firefly algorithm. Section 7.3 describes the procedure for a network model and fitness function formulation technique. The proposed protocol EECFF is presented in Section 7.4. Section 7.5 provides a detailed comparative analysis of LEACH, LEACH-C, and EECFF protocols. Finally, Chapter is summarized in Section 7.6.

7.2 Overview of Firefly Algorithm

Fireflies are extraordinary insects that glow during night by emitting light. This natural phenomenon helps fireflies attract other fireflies for mating purposes and to hunt other insects. Observing the biological behavior of fireflies, Yang proposed a metaheuristic optimization technique called FA [24]. Basically, FA works on following three ground rules:

1. Fireflies emit light to attract other fireflies by emitting light. Fireflies are unisexual and sometimes uses this attraction to hunt other fireflies and insects.
2. The level of attraction depends upon the intensity of the light, i.e., fireflies are attracted to the brightest fireflies.
3. Light intensity is formulated as objective function to be solved for a particular problem.

7.3 Fitness Function Formulation

In any cluster-based routing protocol, CH election is a crucial phase as it decides the fate of the network. Efficient CHs will enhance the network life by keeping the nodes operative for a longer duration and a poor choice of CHs will deteriorate the network. Hence electing efficient CHs is key for energy efficiency. Then, the question becomes "What makes a node efficient enough to be elected as a CH?" To answer this question, we discuss four important parameters, namely node density, cluster fairness, expected energy consumption, and the energy components for which the CHs are elected.

7.3.1 Basic Assumptions and the Network Model

Before the discussion of fitness function formulation, we will discuss some of the basic network assumptions incorporated while simulating the protocols. All the nodes in the simulation are assumed be static, including BS. Each node is equipped with 2 Joules of initial energy. Sensor nodes are deployed in random fashion and grid-based topologies depend upon the network scenario. Each sensor node consumes energy for the activities of data transmission, reception, and processing. A node is assumed to be dead if it is no longer able to transmit or receive data. All nodes are able to communicate with one another over a wireless medium.

For implementing the proposed protocol, the network model mentioned in [10,25] is used. As per this model, for a $node_i$ to transmit $k - bits$ of information over a distance *dist*, it needs $E_t^{(i)}$ energy as shown in Equation 7.1.

$$E_t^{(i)}(k,b) = \begin{cases} k \times Ene_{elec} + k \times \varepsilon_{fs} \times dist^2, & dist < dist_0 \\ k \times Ene_{elec} + k \times \varepsilon_{mp} \times dist^4, & dist \geq dist_0 \end{cases} \tag{7.1}$$

where, Ene_{elec} represents the small amount of energy utilized due to underlying hardware. Ene_{elec} is assumed be fixed for certain hardware types. Energy utilization due to multi-path amplification and the free space channel is represented by ε_{mp} and ε_{fs}, respectively. The transmission distance threshold is indicated by $dist_0$. Similarly, for a $node_i$ to receive $k - bits$ of information, it needs $E_r^{(i)}$ energy, as shown in Equation 7.2.

$$E_r^{(i)} = k \times Ene_{elec} \tag{7.2}$$

7.3.2 Node Density

The node density $\left(N_D^{(i)}\right)$ of a $node_i$ is defined as number of neighbor nodes of $node_i$. It is expected that a CH should be reachable to maximum number of nodes and it should reside at a densely populated area of the network [26]. The CHs with maximum node density result in better network coverage. Hence, one of the objectives of CH election of the presented protocol is to maximize the number of CH nodes with maximum node density. Node density $\left(N_D^{(i)}\right)$ of a $node_i$ is calculated as per Equation 7.3.

$$N_D^{(i)} = \left|nbr^{(i)}\right| \tag{7.3}$$

where, $nbr^{(i)}$ is set of neighbor nodes of $node_i$.

7.3.3 Cluster Fairness

The position of the CH node within the cluster is an important factor. The elected CH node is expected to be relatively closer to all of its cluster members. This ensures that the distance between the cluster member nodes to its CH node is reduced and ultimately intracluster energy consumption is lowered. Cluster fairness $\left(C_f^{(i)}\right)$ of a $node_i$ determines the compactness of the formed cluster. $C_f^{(i)}$ and is calculated per Equation 7.4.

$$C_f^{(i)} = N_D^{(i)} \Big/ \sum_{y \epsilon nbr^{(i)}} dist(x,y) \tag{7.4}$$

where, $dist(x, y)$ represents the Euclidean distance between source node $node_x$ and its neighbor node $\left(node_y \; \epsilon \; nbr^{(i)}\right)$. If the coordinante position of $node_x$ and $node_y$ is assumed to be (α_x, β_x) and (α_y, β_y), respectively, then Euclidean distance between $node_x$ and $node_y$ is calculated per Equation 7.5.

$$dist(x,y) = \sqrt{\left(\alpha_x - \alpha_y\right)^2 + \left(\beta_x - \beta_y\right)^2} \tag{7.5}$$

7.3.4 Expected Energy Consumption

An arbitrary $node_i$ participating in the CH election can estimate the amount of energy to be consumed if it gets elected as CH node through the parameter expected energy consumption $\left(E_{ex}^{(i)}\right)$. It is calculated as per Equation 7.6.

$$E_{ex}^{(i)} = \left[E_t^{(i)}(k, distToBS) + \left(N_D^{(i)} * E_r^{(i)}(k)\right)\right] \tag{7.6}$$

where, $\left(E_t^{(i)}(k, distToBS)\right)$ indicates the energy needed by $node_i$ to send $k - bits$ of message to the base station. $\left(N_D^{(i)} * E_r^{(i)}(k)\right)$ represents amount of energy needed by $node_i$ to receive k-bit data message received from each of the neighbor node $N_D^{(i)}$.

If $node_i$ becomes CH, it is surrounded by $N_D^{(i)}$ number of nodes. Each $node_y \; \epsilon \; nbr^{(i)}$ transmits data to $node_i$ during the allotted TDMA slot. Thus, to receive $N_D^{(i)}$ number of data messages, $node_i$ should consume $\left(N_D^{(i)} * E_r^{(i)}(k)\right)$ Joules of energy. Similarly, once the data is received, the CH node has to send the aggregate information to its BS over a distance *distToBS*. In general, an arbitrary node can estimate the amount energy consumed if it gets elected as CH node. It desirable to elect the node that consumes less energy as a CH node.

7.3.5 Energy Components

In the entire clustering process, the CH is the one that expends comparatively large energy. Hence, it is desirable to elect a node as the CH that has larger residual energy. An arbitrary $node_i$ can estimate its energy component $\left(E_{comp}^{(i)}\right)$ as per Equation 7.7.

$$E_{comp}^{(i)} = E_{curr}^{(i)} / E_{init}^{(i)} \tag{7.7}$$

where, $E_{curr}^{(i)}$ and $E_{init}^{(i)}$ are the residual energy and initial energy of $node_i$, respectively.

To sum up, four important parameters are described so far, namely node density $\left(N_D^{(i)}\right)$, cluster fairness $\left(C_f^{(i)}\right)$, expected energy consumption $\left(E_{ex}^{(i)}\right)$, and energy component $\left(E_{comp}^{(i)}\right)$. Considering the above parameters, CHs are elected after evaluating the fitness function, as shown in Equation 7.8.

$$f^{(i)} = \left(\lambda_1 \times N_D^{(i)}\right) + \left(\lambda_2 \times C_f^{(i)}\right) + \left(\lambda_3 \times E_{ex}^{(i)}\right) + \left(\lambda_4 \times E_{comp}^{(i)}\right) \tag{7.8}$$

where, λ_i $(1 \leq I \leq 4)$ are weights that are set appropriately such that $\lambda_1 + \lambda_2 + \lambda_3 + \lambda_4 = 1$.

7.4 EECFF: Proposed Clustering Protocol

In Chapter 7, an EECFF is presented. The presented protocol is composed of three important phases, namely the cluster head election phase, node association phase, and data transmission phase. Figure 7.1 shows the different phases of the proposed routing scheme.

7.4.1 Cluster Head Election Phase

Cluster head election is the first phase of EECFF protocol. To elect the CHs, we use the firefly algorithm. Let there be n nodes in the network represented as set $\{N_1, N_2, N_3, \ldots, N_n\}$. Out of the n nodes, r nodes are assumed to be elected as cluster heads. Let there be d fireflies and initially each firefly represents set of r arbitrary nodes of the network i.e., $F^{(i)} = \{N_1, N_2,, \ldots, N_r\}$ where $\{N_1 \neq N_2 \neq \ldots \neq N_r\}$ and $f^{(N_1)} \leq f^{(N_2)} \leq \ldots f^{(N_r)}$. Here, each firefly is any random set of nodes and the nodes are arranged in increasing order of their fitness value within each firefly.

After the firefly initialization, each firefly calculates its intensity value. The intensity of a firefly is a hypothetical term, and it actually represents the average fitness value of the nodes that are part of the firefly. The intensity of a firefly F is calculated as per Equation 7.9.

$$Intensity_F^{(i)} = \frac{1}{r}\sum_{j=1}^{r} f^{(j)} \tag{7.9}$$

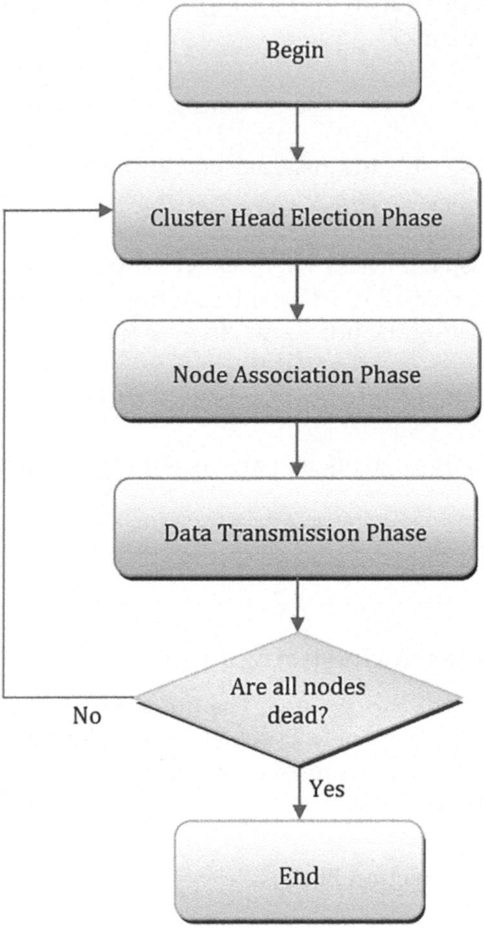

FIGURE 7.1
Phase of EECFF.

As per the basic firefly algorithm [24], fireflies attract each other based on their emitted light intensity. The fireflies that emit high-intensity light attract low-intensity fireflies. Similarly, in our presented protocol, high-intensity fireflies represent the best set of candidate CH nodes. The logic here is to replace the fireflies with low intensity by high-intensity fireflies. In our presented protocol, each firefly with low intensity compares its node set with the node set of the high-intensity firefly and updates its node set with the few best nodes of high-intensity firefly and again evaluates its intensity value. If there is enhancement in intensity, the updated node set is considered. Otherwise, the existing node set is retained. Firefly evaluation and firefly update process are considered as one iteration. The presented protocol continues to perform such iterations till the final set of CHs for the current round are elected.

7.4.2 Node Association Phase

Once the CH nodes are elected, the next task is to associate the rest of the nodes to any one of the available CH nodes. The nodes come to know about CH nodes if they hear a join request message from CH node. Once a node hears a join request message, it means that there is a CH available within its transmission zone. There are possibilities where a node receives two or more join request messages from two or more cluster head nodes. Under such circumstances, a node joins the nearest CH node [27]. It is ideal for any node to join the nearest CH node. But, assigning a node to the nearest CH need not be the only efficient solution. In Chapter 7, we are proposing firefly algorithm–based node association to the CH node, which is clever way of cluster formation.

The node association in the presented protocol is made intelligent using the firefly algorithm. Let there be q number of cluster heads elected for the round m represented as $\{C_1, C_2,\ldots, C_q\}$. Each of the CH nodes broadcast a join request message to the network. Let us assume that a firefly is associated with every CH node. The intensity of a firefly is nothing but the fitness value of the node calculated, as per Equation 7.8. A node may receive multiple join request messages from CHs and nodes can face following two cases while identifying its CH node:

Case 1: A node receives only one join request message. It means that the there is only one accessible CH node. Since there is no option of making choices to choose the CH, the node simply joins the available CH node.

Case 2: A node receives two or more request messages. It means that node is surrounded by two or more CH nodes. In this case, as per the presented node association method, a node chooses a CH node that has a comparatively larger intensity. The presented approach helps in distributing the network load among a CH node efficiently.

7.4.3 Data Transmission Phase

Data transmission is the final phase of the presented protocol and is almost same as the steady state phase of LEACH. In this phase, the elected CH nodes create a TDMA schedule and allot the same to respective cluster members. The cluster member nodes transmit the fetched information to their CHs during the allotted TDMA slot. Finally, CH nodes aggregate the information obtained by all of their members and forward the message to the BS.

The above-mentioned three phases together constitute a single round. Once all the phases are over, a new round commences with a fresh set of CH nodes. This process is iterative and continues till all the nodes are dead.

7.5 Results

The presented protocol is implemented in a network simulator version 2.34 [28]. NS 2.34 inherently does not support WSN routing simulations. We have used MIT uAMPS code extension [29] and made NS2 suitable for our requirements. MIT uAMPS code extension provides LEACH and LEACH-C implementation and the presented protocol, EECFF, is extended from the LEACH source code.

The simulations are carried out by considering varying number of sensor nodes (100, 200, 300, and 400 nodes) with random and grid-based topologies. The network scenarios considered are mentioned below.

- Scenario 1: Randomly distributed nodes across a $(250 \times 250)m^2$ area, BS is situated at the location (125, 125).
- Scenario 2: Nodes arranged in grid topology across a $(250 \times 250)m^2$ area, BS is situated at the location (125, 125).

Simulation parameters considered while implementing the EECFF are mentioned in Table 7.1. EECFF is compared with LEACH and LEACH-C protocols by considering an evaluation metric, such as the number of alive nodes, energy consumption of network, and average number of packets received by the BS. Further subsections describe the comparative analysis of LEACH, LEACH-C, and EECFF protocols.

7.5.1 Number of Alive Nodes

The number of alive nodes represent the node death pattern of the network. During the commencement of simulation, all the nodes are equipped with 2 Joules of battery power. Over the course of the network simulation, the nodes gradually deplete their battery life and soon run out of energy and become inoperable (dead). A good protocol always keeps the network operative for longer duration of time. Figures 7.2 to 7.5 show a comparative plot of

TABLE 7.1

Network Simulation Parameters

Simulation Parameters	Values
Number nodes	100, 200, 300, 400
Simulation area	$(250 \times 250)m^2$
Initial Energy of node	2 Joules
E_{elec}	50 nJ/bit
ε_{fs}	10 pJ/bit/m2
ε_{mp}	0.0013 pJ/bit/m4

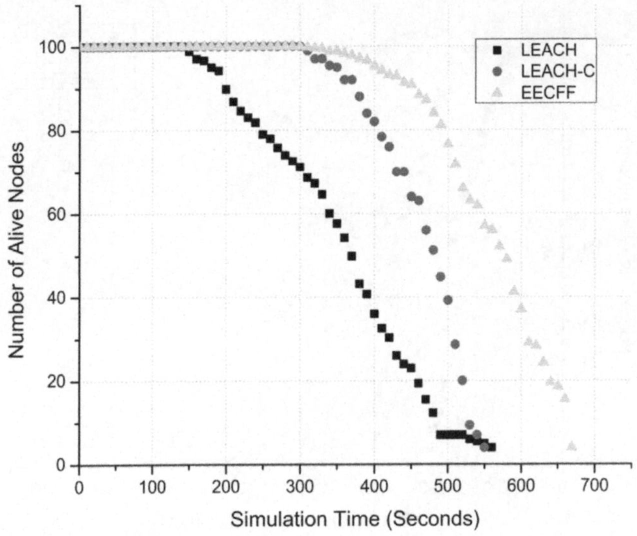

FIGURE 7.2
Number of alive nodes (Scenario 1: 100 nodes).

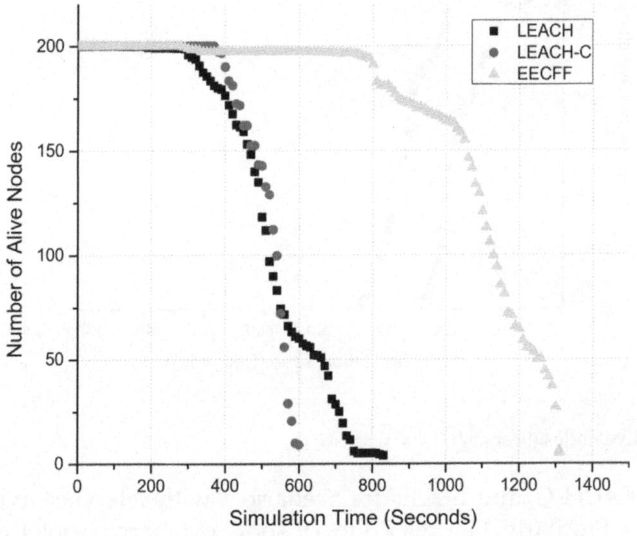

FIGURE 7.3
Number of alive nodes (Scenario 1: 200 nodes).

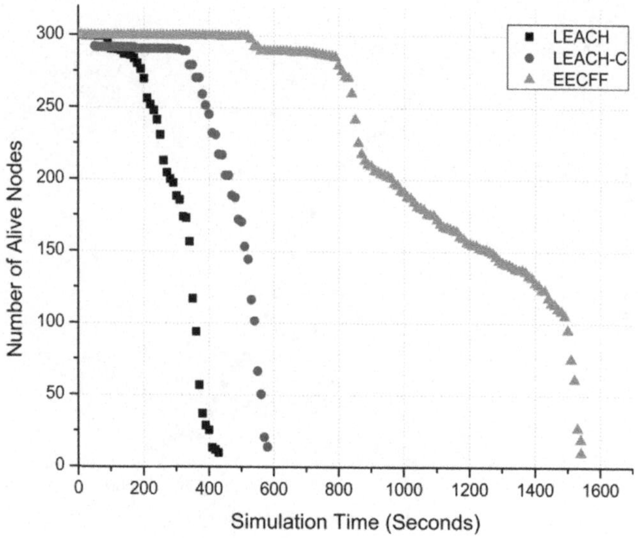

FIGURE 7.4
Number of alive nodes (Scenario 1: 300 nodes).

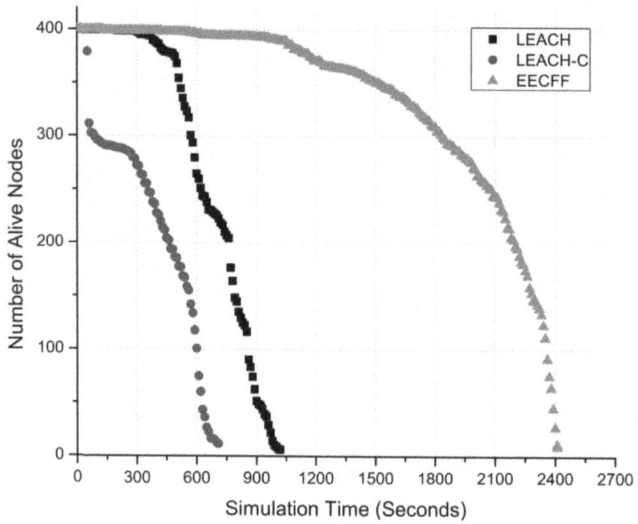

FIGURE 7.5
Number of alive nodes (Scenario 1: 400 nodes).

LEACH, LEACH-C, and EECFF for Scenario 1 with reference to number of alive nodes. Similarly, Figures 7.6 to 7.9 show comparative plot of LEACH, LEACH-C, and EECFF for Scenario 2 with reference to number of alive nodes. It is clear from the figures that, EECFF is able prolong network lifetime better then LEACH and LEACH-C protocols in both the network scenarios.

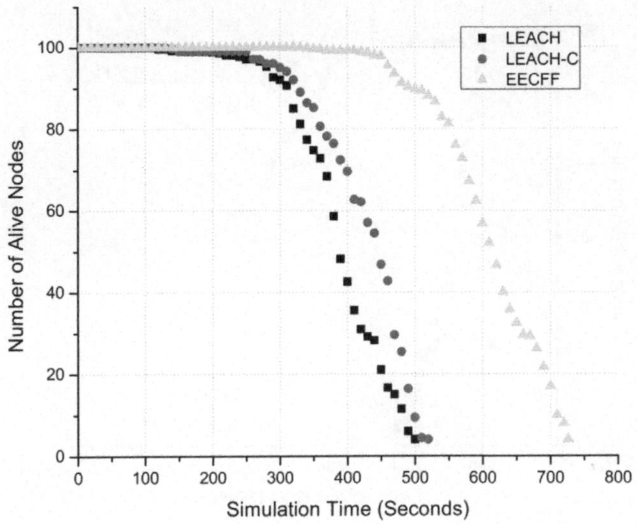

FIGURE 7.6
Number of alive nodes (Scenario 1: 100 nodes).

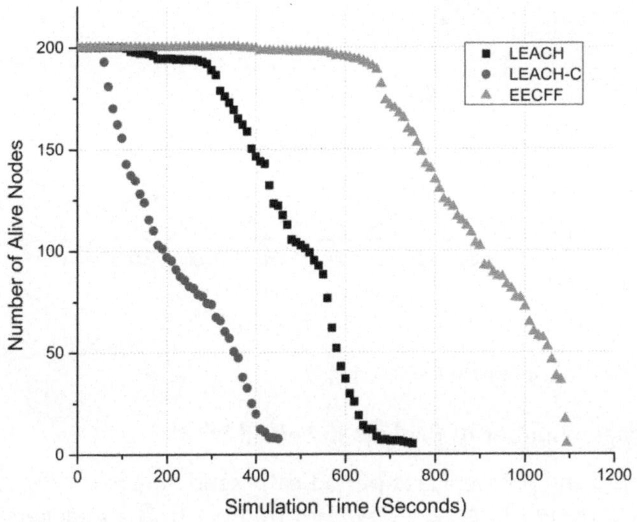

FIGURE 7.7
Number of alive nodes (Scenario 1: 200 nodes).

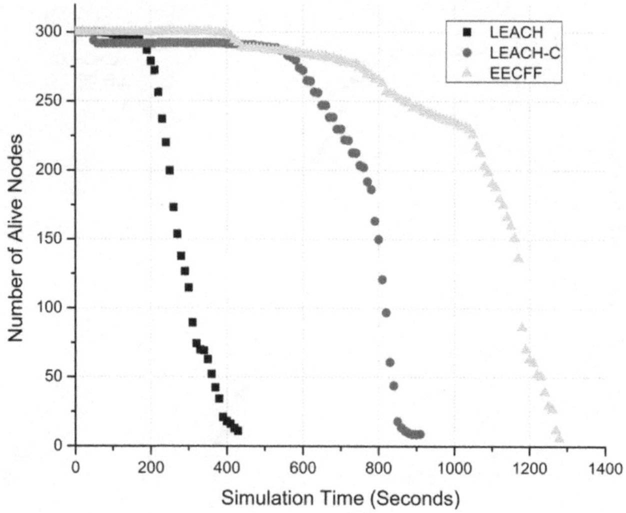

FIGURE 7.8
Number of alive nodes (Scenario 1: 300 nodes).

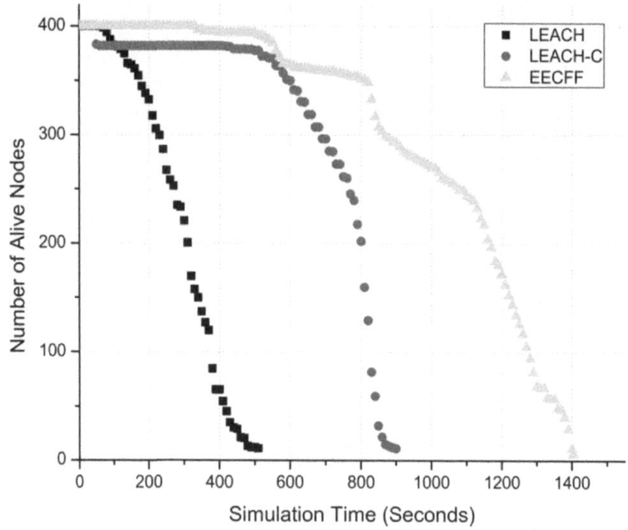

FIGURE 7.9
Number of alive nodes (Scenario 1: 400 nodes).

7.5.2 Average Number of Packets Received by BS

The role of routing protocol is to periodically send data to the BS. An efficient protocol is expected to send maximum number of data packets to the BS. Figures 7.10 and 7.11 show number of data packets received by BS for 100, 200, 300, and 400 nodes cases of Scenario 1 and Scenario 2 respectively.

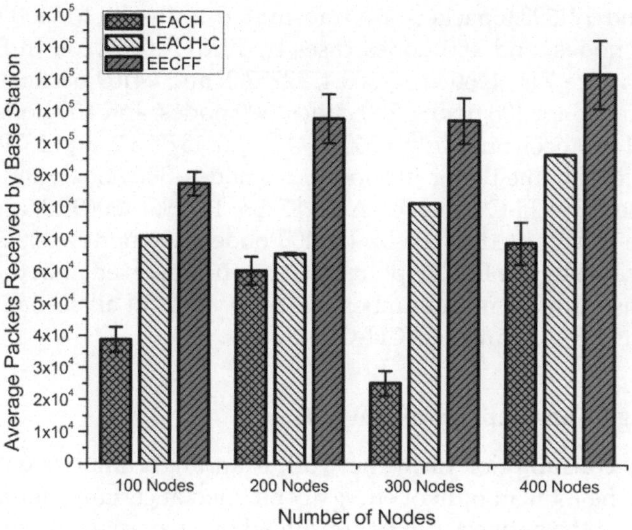

FIGURE 7.10
Average number of packets received by BS (Scenario 1).

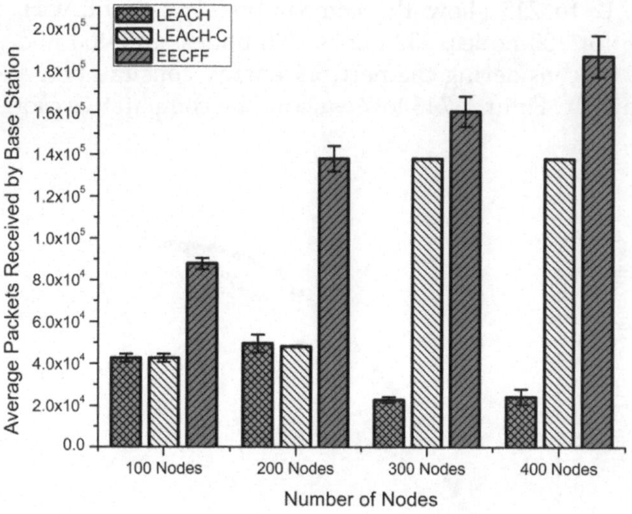

FIGURE 7.11
Average number of packets received by BS (Scenario 2).

In Scenario 1 as shown in Figure 7.10, 38650.8, 60096.4, 25111.6, and 68785 packets are transmitted to the BS for 100 nodes, 200 nodes, 300 nodes, and 400 nodes cases by the LEACH protocol. Similarly, 70908, 65354.8, 81238.6, and 96407.4 packets are transmitted to BS for 100 nodes, 200 nodes, 300 nodes, and 400 nodes cases by the LEACH-C protocol. And, 87188, 107560,

106956.8, and 121393.6 packets are transmitted to the BS for 100 nodes, 200 nodes, 300 nodes, and 400 nodes cases by EECFF protocol. In Scenario 2, shown in Figure 7.11, 42608.6, 49503.4, 22752.2, and 24169 packets are transmitted to the BS for 100 nodes, 200 nodes, 300 nodes, and 400 nodes cases by the LEACH protocol. Similarly, 42608.6, 48112.6, 137976.2, and 137970 packets are transmitted to the BS for 100 nodes, 200 nodes, 300 nodes, and 400 nodes cases by the LEACH-C protocol. And, 87756, 137986, 160695.2, and 187071.2 packets are transmitted to the BS for 100 nodes, 200 nodes, 300 nodes, and 400 nodes cases by the EECFF protocol. It can be observed, that the EECFF is able to transmit maximum number of data packets in all scenarios. EECFF outperforms LEACH and LEACH-C protocols.

7.5.3 Energy Consumption of Network

The energy consumption of the network is an important evaluative metric that shows the total amount of energy consumed at certain simulation time. The energy in the network can be consumed due to various reasons, such as data packet transmission, reception, data processing, etc. An efficient routing protocol should be able to minimize the energy consumption and prolong the network lifetime.

Figures 7.12 to 7.15 show the comparative plot of LEACH, LEACH-C, and EECFF for 100 nodes, 200 nodes, 300 nodes, and 400 nodes cases for Scenario 1, by considering the network energy consumption as evaluative metric. Similarly, Figures 7.16 to 7.19 show the comparative plot of LEACH,

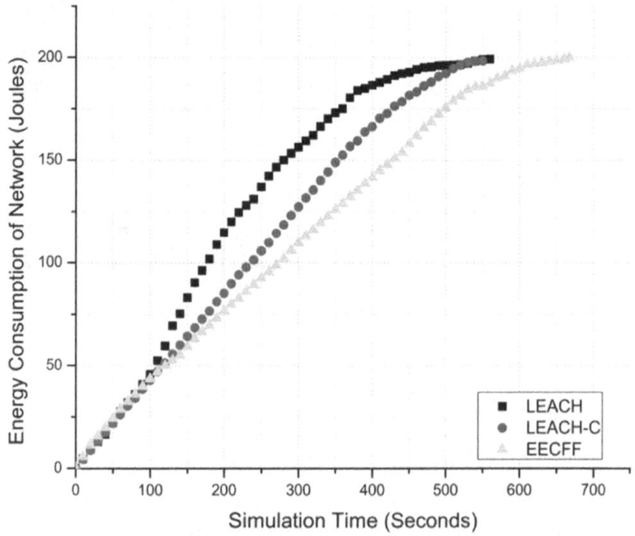

FIGURE 7.12
Energy consumption of network (Scenario 1: 100 nodes).

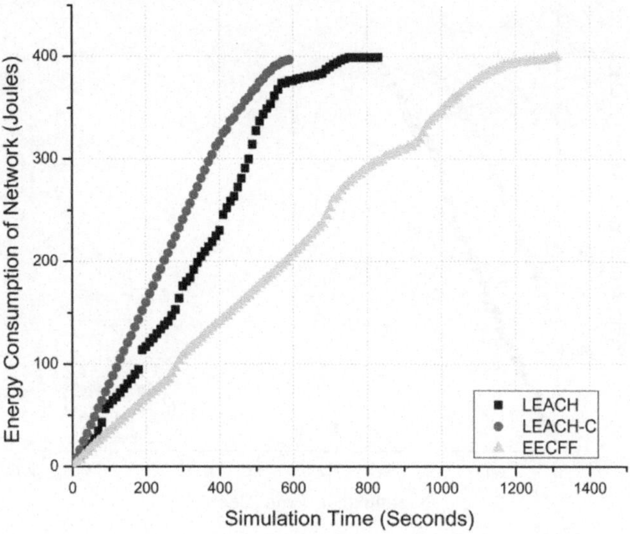

FIGURE 7.13
Energy consumption of network (Scenario 1: 200 nodes).

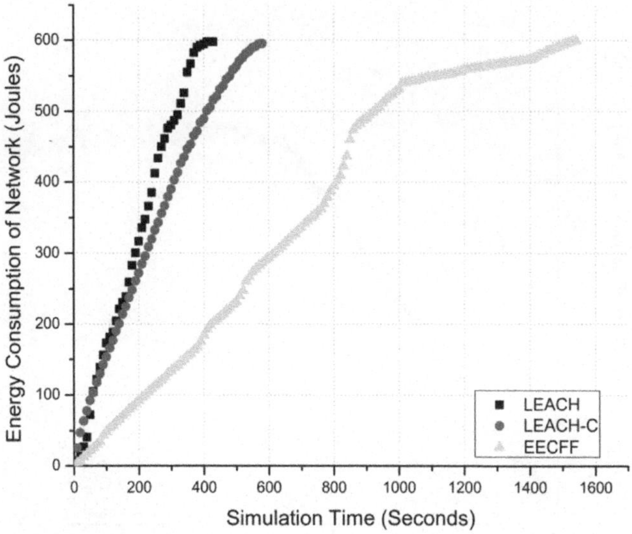

FIGURE 7.14
Energy consumption of network (Scenario 1: 300 nodes).

LEACH-C, and EECFF for 100 nodes, 200 nodes, 300 nodes, and 400 nodes cases for Scenario 2, by considering network energy consumption as evaluative metric. It can be observed from the graphs that, EECFF is able to effectively utilize the network energy and is way better than LEACH and LEACH-C protocols.

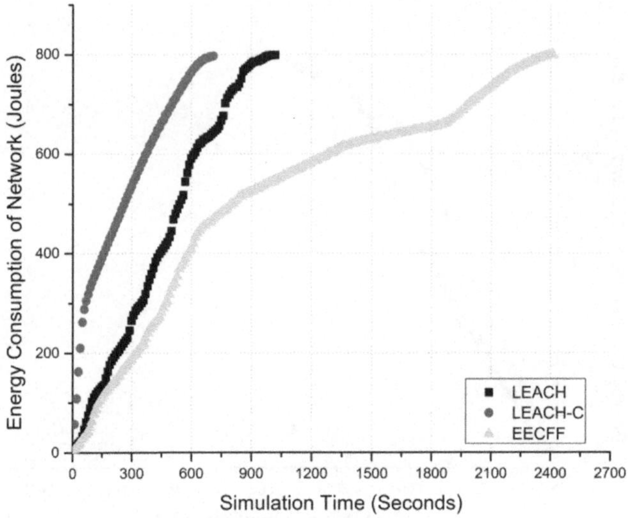

FIGURE 7.15
Energy consumption of network (Scenario 1: 400 nodes).

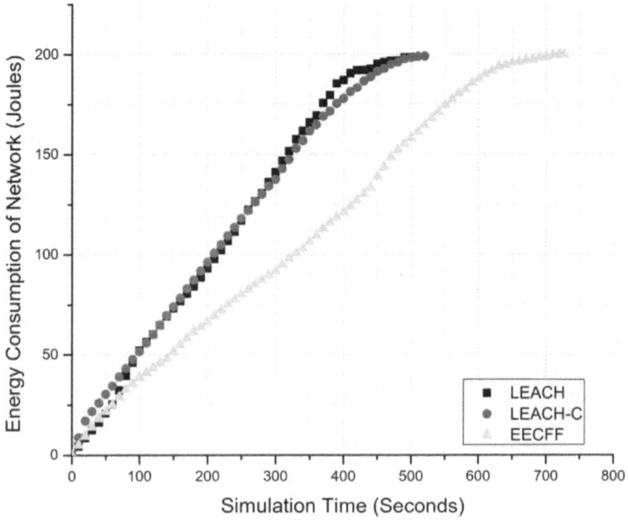

FIGURE 7.16
Energy consumption of network (Scenario 2: 100 nodes).

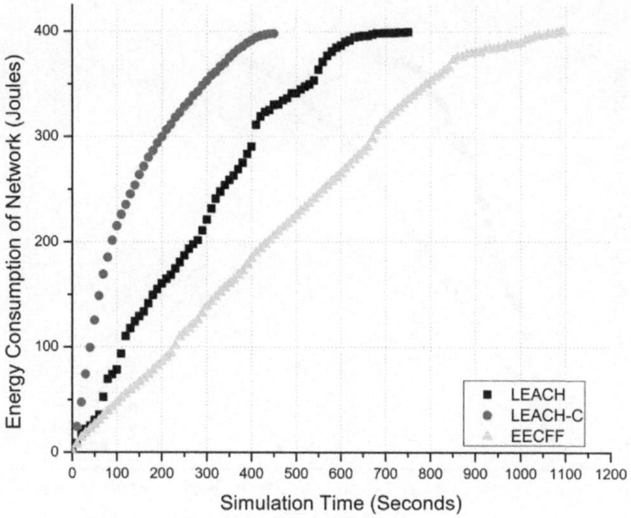

FIGURE 7.17
Energy consumption of network (Scenario 2: 200 nodes).

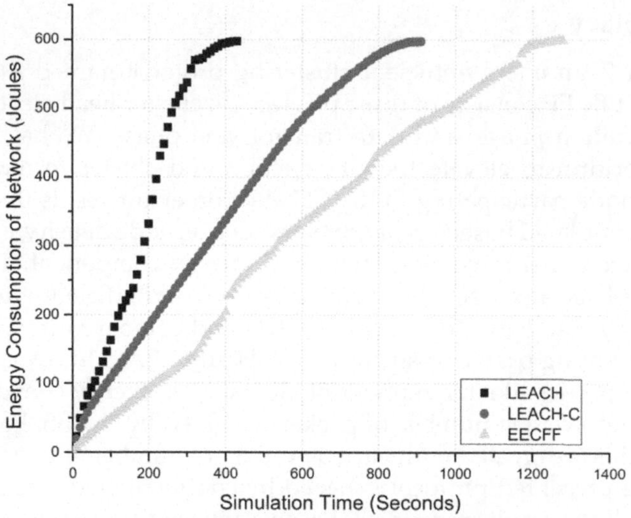

FIGURE 7.18
Energy consumption of network (Scenario 2: 300 nodes).

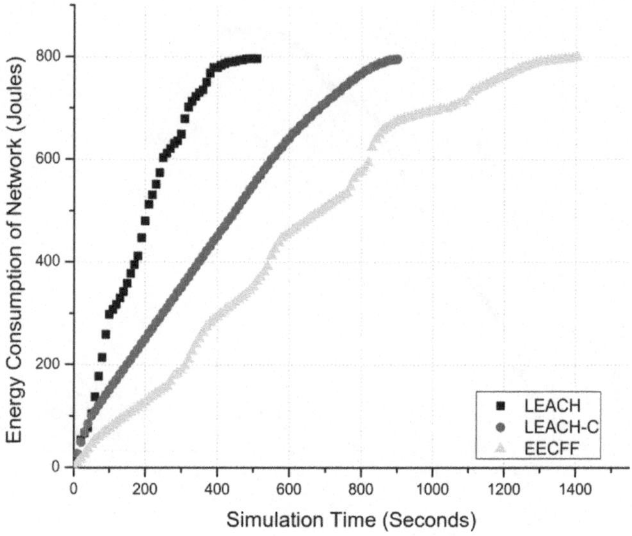

FIGURE 7.19
Energy consumption of network (Scenario 2: 400 nodes).

7.6 Summary

In Chapter 7, an energy-efficient clustering protocol named as EECFF is presented. EECFF consists of three phases: the cluster head election phase, node association phase, and data transmission phase. We have used the firefly algorithm to elevate the CH election and cluster formation strategy. Each node participating in the CH election evaluates its fitness value, which is formulated based on parameters such as node density, cluster fairness, expected energy consumption, and energy component. The cluster formation phase also uses the firefly algorithm and efficiently forms clusters. EECFF is implemented in NS ver. 2.34 and compared it with some of the legacy routing protocols such as LEACH and LEACH-C by considering evaluation metrics like the number of alive nodes, energy consumption of network, and average number of packets received by the BS. For analysis, two scenarios are created with varied number of number of nodes from 100 to 400. The presented protocol is tested by conducting several simulation trials and all the results plotted are based on the average values of the multiple simulation trials. From the obtained results, it is clearly observed that EECFF performs efficiently in both random and grid-based topologies. The reason for the improved results is due to efficient cluster head and cluster formation strategies.

References

1. Kazem Sohraby, Daniel Minoli, and Taieb Znati. Wireless Sensor Networks: Technology, Protocols, and Applications. *Wiley-Interscience,* 2007.
2. Dey N, Ashour AS, Shi F, Sherratt RS. Wireless Capsule Gastrointestinal Endoscopy: Direction-of-Arrival Estimation Based Localization Survey. *IEEE Reviews in Biomedical Engineering,* 10:2–11, 2017.
3. Dey N, Ashour AS, Shi F, Fong SJ, Sherratt RS. Developing residential wireless sensor networks for ECG healthcare monitoring. *IEEE Transactions on Consumer Electronics,* 63(4):442–449, November 2017.
4. Jun Zheng and Abbas Jamalipour. *Wireless Sensor Networks: A Networking Perspective.* Wiley-IEEE Press, 2009.
5. Akkaya Kemal and Mohamed Younis. A survey on routing protocols for wireless sensor networks. *Ad Hoc Networks,* 3(3):325–349, 2005.
6. N. A. Pantazis, S. A. Nikolidakis, and D. D. Vergados. Energy efficient routing protocols in wireless sensor networks: A survey. *IEEE Communications Surveys Tutorials,* 15(2):551–591, Second 2013.
7. Wenjing Guo and Wei Zhang. A survey on intelligent routing protocols in wireless sensor networks. *Journal of Network and Computer Applications,* 38:185, 201, 2014.
8. Tifenn Rault, Abdelmadjid Bouabdallah, and Yacine Challal. Energy efficiency in wireless sensor networks: A top-down survey. *Computer Networks,* 67:104–122, 2014.
9. S. K. Singh, P. Kumar, and J. P. Singh. A Survey on Successors of LEACH Protocol. *IEEE Access,* 5:4298–4328, 2017.
10. Wendi Rabiner Heinzelman, Anantha Chandrakasan, and Hari Balakrishnan. Energy efficient communication protocol for wireless microsensor networks. *Proceedings of the 33rd Hawaii International Conference on System Sciences* Volume 8, HICSS '00, 8020, Washington, DC, USA, 2000. IEEE Computer Society.
11. Arati Manjeshwar and Dharma P. Agrawal. Teen: A routing protocol for enhanced efficiency in wireless sensor networks. *Proceedings of the 15th International Parallel & Distributed processing Symposium, IPDPS '01,* 189, Washington, DC, USA, 2001. IEEE Computer Society, pp. 2009–2015.
12. S. Lindsey and C. S. Raghavendra. Pegasis: Power-e-cient gathering in sensor information systems. In Proceedings, *IEEE Aerospace Conference,* 3, 1125–1130, 2002.
13. O. Younis and S. Fahmy. Heed: a hybrid, energy efficient, distributed clustering approach for ad hoc sensor networks. *IEEE Transactions on Mobile Computing,* 3(4):366–379, Oct. 2004.
14. Muhammad Saleem, Gianni A. Di Caro, and Muddassar Farooq. Swarm intelligence based routing protocol for wireless sensor networks: Survey and future directions. Information Sciences, 181(20):4597–4624, 2011. *Special Issue on Interpretable Fuzzy Systems.*
15. A. M. Shamsan Saleh, B. M. Ali, M. F. A. Rasid, and A. Ismail. A survey on energy awareness mechanisms in routing protocols for wireless sensor networks using optimization methods. *Transactions on Emerging Telecommunications Technologies,* 25(12):1184–1207, 2014.

16. Adamu Murtala Zungeru, Li-Minn Ang, and Kah Phooi Seng. Classical and swarm intelligence based routing protocols for wireless sensor networks: A survey and comparison. *Journal of Network and Computer Applications*, 35(5):1508–1536, 2012. Service Delivery Management in Broadband Networks.
17. Binh HT, Hanh NT, Dey N. Improved cuckoo search and chaotic flower pollination optimization algorithm for maximizing area coverage in wireless sensor networks. *Neural Computing and Applications*, 1–3, 2016.
18. Vijayalakshmi K, Anandan P. A multi objective Tabu particle swarm optimization for effective cluster head selection in WSN. *Cluster Computing*, 1–8, 2018.
19. Dey N, Samanta S, Chakraborty S, Das A, Chaudhuri SS, Suri JS. Firefly algorithm for optimization of scaling factors during embedding of manifold medical information: an application in ophthalmology imaging. *Journal of Medical Imaging and Health Informatics*, 4(3):384–94, June 2014.
20. Kumar R, Talukdar FA, Dey N, Balas VE. Quality Factor Optimization of Spiral Inductor using Firefly Algorithm and its Application in Amplifier. *International Journal of Advanced Intelligence Paradigms*, (In-press).
21. Chakraborty S, Dey N, Samanta S, Ashour AS, Balas VE. Firefly algorithm for optimized nonrigid demons registration. *Bio-inspired Computation and Applications in Image Processing*, 221–237, 2017.
22. Jagatheesan K, Anand B, Samanta S, Dey N, Ashour AS, Balas VE. Design of a proportional-integral-derivative controller for an automatic generation control of multi-area power thermal systems using firefly algorithm, *IEEE/CAA Journal of Automatica Sinica*, January 25, 2017.
23. Kumar R, Rajan A, Talukdar FA, Dey N, Santhi V, Balas VE. Optimization of 5.5-GHz CMOS LNA parameters using firefly algorithm. *Neural Computing and Applications*, 28(12):3765–3779, December 2017.
24. Xin-She Yang. Firefly algorithms for multimodal optimization. *Proceedings of the 5th International Conference on Stochastic Algorithms: Foundations and Applications*, SAGA'09, pages 169–178, Berlin, Heidelberg, 2009. Springer-Verlag.
25. Wendi Rabiner Heinzelman, Anantha Chandrakasan, and Hari Balakrishnan. An application-specific protocol architecture for wireless microsensor networks. *Trans. Wireless. Comm.*, 1(4):660–670, October 2002.
26. Buddha Singh and Daya Krishan Lobiyal. A novel energy-aware cluster head selection based on particle swarm optimization for wireless sensor networks. *Human-centric Computing and Information Sciences*, 2(1):13, 2012.
27. Pratyay Kuila and Prasanta K. Jana. Energy efficient clustering and routing algorithms for wireless sensor networks: Particle swarm optimization approach. *Engineering Applications of Artificial Intelligence*, 33:127–140, 2014.
28. Network Simulator Ver. 2, Available Online at. https://www.isi.edu/nsnam/ns/.
29. MIT uAMPS LEACH ns Extensions, Available Online at. http://www.ece.rochester.edu/projects/wcng/code.html.

8

Positioning Improvement of Sensors in Wireless Sensor Networks

N. Pushpalatha, K. Ramani, and B. Anuradha

CONTENTS

8.1 Overview of Wireless Sensor Networks

Wireless sensor networks (WSNs) are networks of distributed, autonomous nodes that can sense or monitor physical or environmental conditions cooperatively [1]. Each sensor node consists of one or more sensors, a radio transceiver, a microcontroller, and an energy source. Due to their potential applications in many areas ranging from environmental observation, natural habitat monitoring, medicine, industry, and the military, WSNs have attracted a lot of interest in recent years [2].

With the intention of maximizing capabilities like coverage area and the monitoring of sensor networks, a novel deployment strategy should be provided [3]. The deployment of mobile sensor nodes in the region of interest (ROI), where interesting events might happen and the corresponding detection mechanism is required, is one of the key issues in this area [4]. Sensor deployment strategies play a very important role in providing better quality of service, which relates to the issue of how well each point in the sensing field is covered [5]. The most effective approach in sensor deployment is to place sensors in such a manner so that the maximal network coverage is achieved [6]. An efficient deployment of sensor nodes will reduce the construction and communication cost of the network and improve resource management [7]. The deployment process is done according to the following constraints; (i) deployment expenses based on sensor number; (ii) the probability of event detection; (iii) lifetime energy consumption [8]; and (iv) connectivity. Static and the dynamic deployment are two categories into which wireless sensor nodes can be subdivided. Static deployment involves two methods: deterministic and random. In deterministic deployment, sensor nodes are statically deployed, whereas in dynamic deployment, each sensor node moves automatically to a proper location before initiating its work [9].

In general, energy consumption is another major problem in addition to the coverage problem in randomly deployed sensor networks. In many applications, a sensor lifetime is crucial, which affects the effectiveness of mobile sensor networks, since the mobility of sensors consumes high energy rather than processing it. Hence, if once the battery's power runs out of power, then sensor node becomes unusable, and this may degrade the quality of service for the entire wireless sensor network.

8.2 Background

Recently, researchers have witnessed the successful real-world deployments of wireless sensor networks (WSNs) in a wide range of civil and military applications. Wide research is taking place in many areas related to wireless sensor networks, like sensor positioning, communication, coverage, etc. Even personal health monitoring systems are based on wireless technology [10]. Similarly, to overcome the drawbacks of traditional gastrointestinal procedures, wireless capsule-based endoscopy is evolving in the medical field [11]. Chapter 8 especially focuses on the sensor positioning and coverage area of wireless sensor networks. Enhancing the entire coverage area of the total sensor network is one of the main objectives of this area coverage problem. In addition, this area coverage problem influences the performance of applications, specifically target detection and tracking, personal protection, homeland security, monitoring the battlefield, and animal habitat monitoring. This position of sensor nodes is one among the crucial factors that influences network coverage. In order to achieve a better coverage area, the sensors are placed in position deterministically. But, owing to some conditions where sensors can't be positioned deterministically (for instance, in areas of application involving nature calamities or say turbulent environments), there is a need to develop a sensor-positioning technique that meets the maximum criteria and efficient deployment strategies. This should achieve the given things: better connectivity, maximal network coverage, lowered expense on construction and communication, better event detection probability, lowered energy consumption, accurate positioning, reduced position estimation error, and reduced complexity.

In reference to the literature review, a technique based on the glowworm swarm optimization (GSO) [7] has solely taken the distance covered in between the sensor nodes into consideration, while the artificial bee colony (ABC) algorithm [12] has solely taken the coverage rate into consideration. These techniques, in spite of providing accurate positioning, do not agree with some of the other parameters. Received signal strength (RSS)–based positioning techniques [13–15] provide lesser complexity, however they do not ensure cost and energy reduction. Geometric techniques, such as triangulation techniques, use the angle of arrival from various sensors to determine their location. The angle between anchor sensor node and unknown sensor node is calculated in a given reference frame using trigonometry sine and cosine laws. This computation is well briefed [16,17]. But, this technique has its limitations, such as a lower coverage area and higher range error. The sensor positioning technique [18] provides greater accuracy and minimizes the expense, but fails in the areas of maximum coverage and connectivity.

Celal Ozturk et al. [12] made use of the ABC algorithm for the dynamic deployment of stationary and mobile sensor networks, whereas, GSO was applied by Wen-Hwa Liao et al. [7]. Effective computation of the covered area was implemented by making use of the probabilistic detection model which is considered to obtain more realistic results. Optimization techiniques were used with the aim that the coverage rate of the network was maximized by effectively estimating the coverage ratio.

Naveed Salman et al. [15] improved the performance of the RSS-based localization technique. In this technique, he proposed the weighted least squares (WLS) algorithm to improve the location estimation accuracy. To improve the performance of WLS algorithm further, a reference anchor optimization using the minimization of the theoretical mean square error (MSE) was also proposed.

Quan Liu et al. [19] proposed a three-dimensional (3D) accurate positioning algorithm which makes use of the received signal strength indicator (RSSI)–based positioning technique which constructed an attenuation model of wireless signal between unknown node and the beacon node in 3D space. This method compensates for the loss of the wireless signal on the basis of their material, thickness, and other factors when there are obstacles or walls. Positioning accuracy was improved with the help of the error revision algorithm.

A new sensor positioning scheme in nonuniform wireless sensor networks to ensure high accuracy of sensor positioning was proposed by Junho Park et al. [18]. This scheme made use of minimum anchor nodes while their distance was measured w.r.t the neighbor density values in a nonuniform sensor network environment. This method generally utilizes a minimum four anchor nodes, which are positioned at the sensing fields boundary to lower the construction expenses of the sensor network. Tian He et al. [20] have presented the triangulation geometric technique, which uses the angle of arrival to determine the location of all sensors. In this method, the limitations are low coverage area and high range error.

Yun Wang et al. introduced three basic problems like sensing coverage, network connectivity, and sensor placement/deployment in a WSN [21]. Recent years have witnessed the successful real-world deployments of WSNs in a wide range of civil and military applications.

Howard et al. used a technique based on the maximum likelihood to estimate the location of mobile robot [22]. Roumeliotis et al. presented a method based on a distributed Kalman filter to achieve cooperative localization [23]. Fox et al. developed a probabilistic collaborative localization method for location estimation [24].

In order to estimate the signal power at receiver end, the RSSI technique is applied. But, the technique results in relatively low accuracy. However, due to the simplicity of it, this RSSI is adopted widely in many research areas. These parameters, such as time of arrival (ToA) and time difference of arrival (TDoA), were used by Savvides et al. [25,26] and Priyantha et al. [27] to reduce the errors of range estimation, but these methods require equipping each sensor node with a powerful capability of computation. Recently, Niculescu et al. used the angle of arrival (AoA) method to measure the sensors' positions [28]. AoA sensing needs each sensor node to be installed with an antenna array or ultrasound receivers.

A large number of sensor nodes with known positions are densely distributed in a sensor network [29–32]. These nodes with positions known are also called "beacons" or "anchor nodes," their areas range in a grid across the network to estimate other nodes' positions close by.

The distance vector exchanges to estimate the distance from the non-anchor nodes to the anchor nodes. Based on these distances, each node can estimate its position by performing a trilateration or multilateration [31,32]. The performance of the algorithms is deteriorated by the range estimation errors and the inaccurate estimation of distances, that are caused by complex terrain and the anisotropic topology of the sensor network. A node at a position which is not known receives low accuracy range measurements from a large number (>3) of neighboring anchor nodes. Making use of a least-mean squares approach in order to solve the overdefined triangulation results in a solution with an accuracy that is substantially greater than what is expected from the unreliable measurement ranges. Savarese [16] tried to improve the above approach by iterative computing. However, this method added a great deal of communication cost of the algorithm and could not generate a good position estimate in some circumstances. Moreover, the accuracy of this class of algorithms relies on the average radio range estimation, and tends to deteriorate when the topology of a sensor network is anisotropic.

Recently, there has been some research on the error characteristics of sensor localization [33,34] and computation complexity [33]. Chintalapudi et al. [35,36] analyzed factors that impacted the performance of the system and then proposed ad hoc localization systems with ranging information [37]. Eren applied graph rigidity theory to locate sensors. Range constraints [36] and area constraints are used to locate sensors in coarse granularity as well.

Hua Chen et al. [38] made use of the energy efficiency–based maximal lifetime routing algorithms in wireless sensor networks. Gyöngyi Kocsisné Szilágyi et al. [39] proposed measurements of distances between sensor nodes in wireless sensor networks. M.B. Nirmala and A.S. Manjunatha [40] introduced the development of location estimation algorithms with improvement in location with lowered expense.

H.T.T. Binh et al. [41] proposed algorithms, namely improved cuckoo search and chaotic flower pollination, to enhance the area of coverage of sensor nodes with reduced computation time and improved stability.

8.3 Wireless Sensor Network Model

A sensor node is comprised of four fundamental components: a sensing component, processing unit, power unit, and transceiver component, as detailed in Figure 8.1 below. It also has other extra components based on application

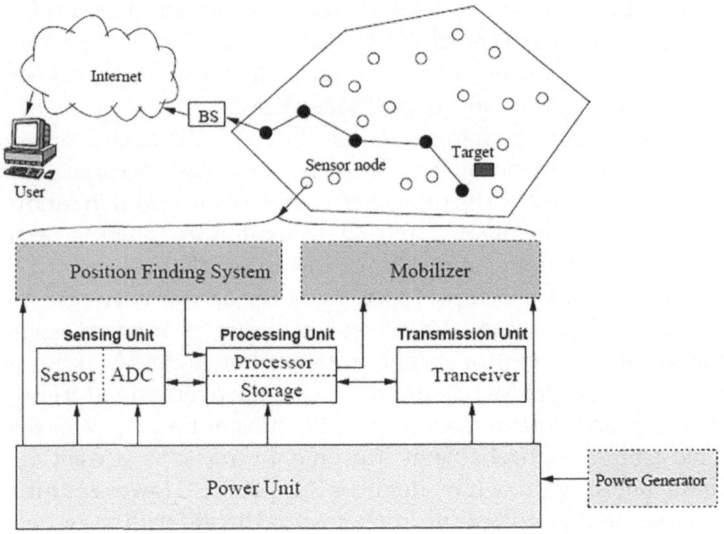

FIGURE 8.1
Block diagram of wireless sensor network.

dependence, like a locating system, a power generator, and a mobilizer. Sensing components are usually comprised of two subunits: sensors and ADCs. The signals (analog) emitted by the sensors are converted to digital signals by the ADC and are given to the processing unit. The processing unit in general is associated with the storage unit, which is small and can look into the processess that makes the sensor node work with the other nodes to carry out the assigned sensing tasks. A transceiver unit connects the node with a network. The power unit is one of the most important units. These power units are given support by means of a power scavenging unit such as solar cells. Additional node subunits are application specific.

The block diagram of the versatile wireless sensing node is depicted in Figure 8.2. The approach of modular design provides a flexible and versatile platform to address the requirements of a wide variety of applications. For instance, based on the deployed sensors, the signal conditioning block can be replaced or reprogrammed. This gives space for a wide range of various sensors to be used along with the wireless sensing node. In the same way, the radio link may be interchanged as needed for any given applications' wireless range requirements and the necessity of bidirectional communication

8.3.1 Position Estimation Elements

Like the very recent recreation of the age-old problem of position estimation, position estimation in sensor networks have drawn the attention of a large

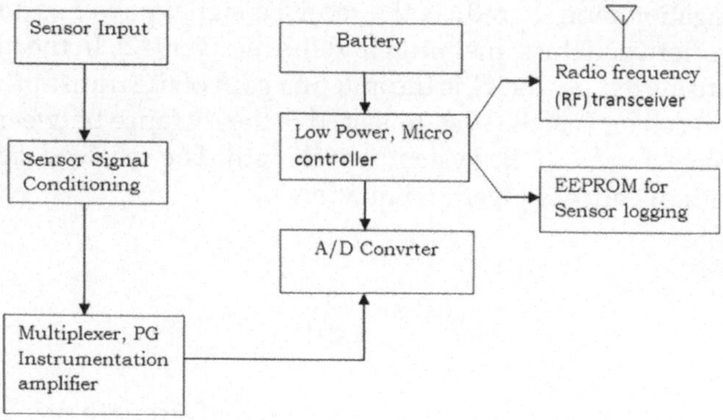

FIGURE 8.2
Sensor node functional block diagram.

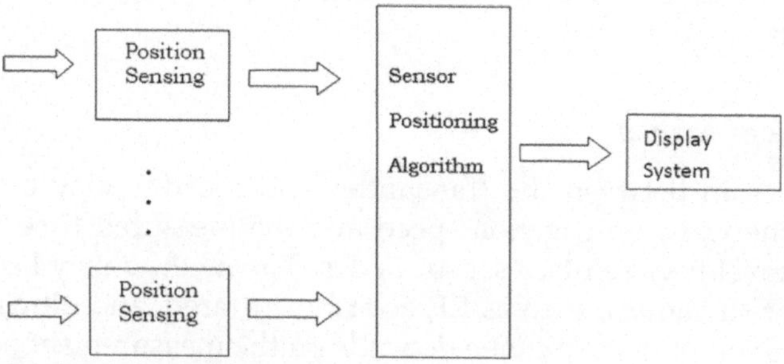

FIGURE 8.3
Positioning sensor block diagram.

research effort in the past ten years. Sensor positioning in sensor network setting is useful for different purposes and is depicted in Figure 8.3.

8.3.2 Received Signal Strength Indication

During radio propagation, an important characteristic is that the radio signal attenuates as the distance between the transmitter and receiver increases. The power of the received radio signal falls off exponentially with distance increasing, and the receiver can measure this attenuation based on RSSI in order to estimate the distance to the sender [42]. RSSI measures the power of the signal at the receiver. Based on the transmit power, the propagation loss is calculated and the loss can be translated into distance estimate. This method has been used mainly for radio frequency (RF) signals.

The radio propagation models are well researched, and they are used to predict the average RSSI at a given distance from the transmitter. An ideal radio propagation model predicts the received signal power as a function of the distance between the transmitter and the receiver [42]. In the ideal model, P_λ is the transmitted power, G_t is the antenna gain of the transmitter, G_r is the gain of the receiver, L is the system loss, d is the distance between transmitter and receiver, and λ is the system wavelength. The received signal power strength with distance is given in Equation 8.1.

$$P_r(d) = \frac{P_\lambda G_t G_r \lambda^2}{4\pi^2 d^2 L}. \tag{8.1}$$

The distance estimation with received RF signal strength using the WINS sensor nodes is studied [25]. In the experiments, different configuration strategies, including different power levels in transmitters and deployment strategies of sensors, are used to estimate the relation between received signal strength and also to estimate the distance between receiver and transmitter units.

8.3.3 Time of Arrival

The distance in between the transmitter and receiver may be estimated based on the wave propagation speed and the measured time for a radio signal to travel between two sensor nodes. The method may be applied to many different signals, such as RF, acoustic, infrared, and ultrasound. The implementation of the technique depends on the measurement of ToA. ToA offers a high level of accuracy, but also requires relatively fast processing capabilities in sensor nodes to resolve many time differences for fine-grained measurements.

8.3.4 Time Difference of Arrival

The distance from transmitter to receiver may be measured by the TDoA of different communication media at different speeds. For example, the measurement for ToA is made based on two different modalities of communication (ultrasound and radio) in sensor nodes. The propagation speeds for ultrasound and radio are considerably different.

8.3.5 Angle of Arrival

AoA means the angle at which signals are received by the receiver from the transmitter. An AoA system is able to estimate the angle at which signals are received and to use simple geometric relationships to estimate the relative locations of transmitter and receiver. AoAs may also be combined with distance estimates to derive relative locations. There are two major disadvantages of the AoA techniques which make it inapplicable to sensor networks. The first is that the cost of the complex antenna array is high. Second, the AoA techniques will not scale well for systems with a large number of nodes.

8.3.6 Triangulation

Triangulation is a geometric technique that uses the angles of arrival to determine the location of sensors. With the angle of each anchor sensor, with respect to the unknown sensor node in some reference frame, the unknown sensor node's locations are calculated with the trigonometry laws of sines and cosines. The computation of triangulation [43] is illustrated by Figure 8.4.

8.3.7 Trilateration

Trilateration is a geometric technique that uses distances between three anchor sensors and one unknown sensor to determine the unknown sensor's location. An unknown sensor is uniquely located when at least three reference points are associated with it in a two-dimensional (2D) space. The location of the unknown sensor is estimated by calculating the intersection of three circles. Figure 8.5 illustrates the computation geometric constraint [25].

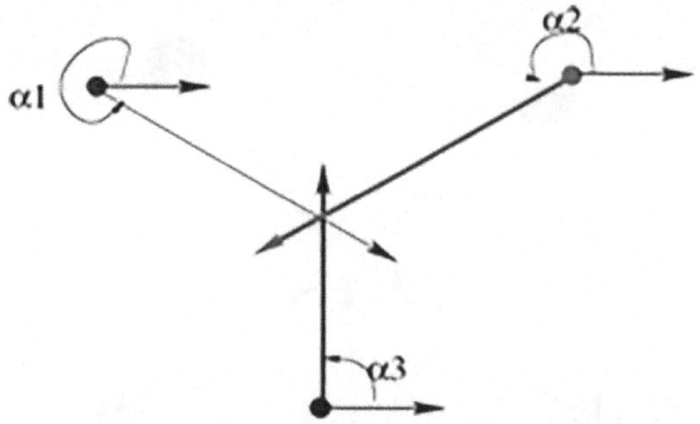

FIGURE 8.4
Triangulation.

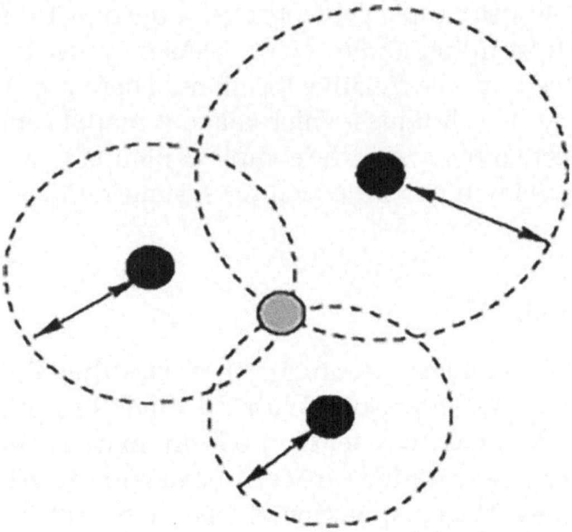

FIGURE 8.5
Trilateration.

8.3.8 Multilateration

An unknown sensor's location may also be estimated with multilateration with its distances to more than three anchor sensors. Beutel studied the multilateration with the least square algorithm [44]. Figure 8.6 illustrates the computation geometry constraint [25].

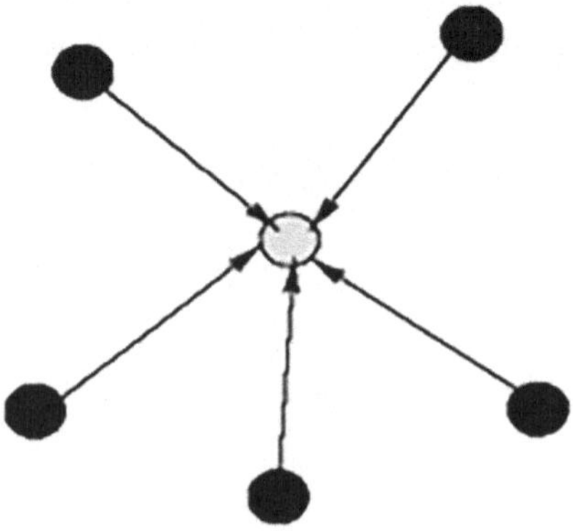

FIGURE 8.6
Multilateration.

For n anchor sensors in 3D space and their distances to the unknown sensor, sensor node can be defined as in Equation 8.2:

$$\begin{bmatrix} d_1^2 \\ d_2^2 \\ . \\ . \\ . \\ d_n^2 \end{bmatrix} = \begin{bmatrix} (x_1 - u_x)^2 + (y_1 - u_y)^2 \\ (x_2 - u_x)^2 + (y_2 - u_y)^2 \\ . \\ . \\ . \\ (x_n - u_x)^2 + (y_1 - u_y)^2 \end{bmatrix} \tag{8.2}$$

where d_i is the distance between the ith anchor sensor and the unknown sensor, (x_i, y_i) is the location of ith anchor sensor in two-dimensional space, and (u_x, u_y) is the location of unknown sensor in 2D space.

8.3.9 Applications of Sensor Networks

By convention, sensor networks were found useful in areas which involved high-end applications like as radiation and nuclear-threat detection systems, over the horizon weapon sensors for ships, habitat sensing, biomedical applications, and seismic monitoring. In short, the focus is on networked biological and chemical sensors for national security applications. And to add more, the evolving interest extends to direct consumer applications [19]. Existing and potential applications of sensor networks consists, among a few others, physical security, military sensing, traffic surveillance, air traffic control, video surveillance, process control, industrial and manufacturing automation, distributed robotics, inventory management, weather sensing, environment monitoring, national border monitoring, and building and structure monitoring. A list of applications follows [45]: military applications, environmental applications, health applications, home applications, and commercial applications.

8.4 Sensor Distribution Strategies

Sensor distribution strategies explore the idea of using dimensionality reduction techniques to estimate the position of sensors in 2D or 3D space. In Chapter 8, the implementation of various methods of wireless sensor networks using multidimensional scaling for sensor positioning technique is presented.

To analyze the variation of data on a set of objects, and to discover the spatial structures in the data [46] the multidimensional scaling (MDS) technique is used. This technique can be used as a data analytic approach to find out the dimensions that lie beneath the judgments of distance and model data in a geometric space. The major benefit in using the MDS for position estimation is that it can generate position estimation with relatively high accuracy, provided that it is based on limited and error-prone distance information. There are many varieties of MDS that exist. On 2D MDS and the iterative optimization of MDS, the fundamental idea is to presume that the variation of data is distance and then infer their coordinates. Additional particulars concerning broad and innate explanation of MDS are obtainable in [47]. Motivated by the above multidimensional scaling techniques, a multivariate optimization–based iterative algorithm for sensor position calculation is presented.

8.4.1 Multidimensional Scaling Algorithm Techniques

With the intention of calculating all sensors' approximate positions in a distributed wireless sensor network, a small percentage of sensors cover their position information known either through manual configuration or equipped with GPS. These sensors with given position information are called "anchor sensors" and other sensors without position information are defined as "unknown sensors." Suppose that all sensors' positions, with the assistance of anchor sensors, are to be determined. In general, the anchor sensors broadcast their positions to their neighbors. The neighboring unknown sensors make use of the broadcasted anchor sensor positions to measure their spatial relation from their neighbors to determine their own positions. An unknown sensor becomes an anchor sensor once after estimating its own position with the help of neighboring sensor information which, in turn, helps in estimating the position of other unknown sensors.

Multidimensional scaling as a technique for WSN localization can be implemented as follows: (i) Generate distance matrix by calculating the distances between every pair of nodes in the network. (ii) Consider the above step output as input for computing first largest eigenvalues and eigenvectors, which gives a relative map with relative location for each given node. (iii) Generate absolute map from relative map with sufficient number of anchor nodes, and (iv) 2D multidimensional scaling: If the distance between pair of sensors is compatible, then 2D multidimensional scaling can be applied. (v) 2D iterative multidimensional scaling: When the distance between certain pairs of sensors is unknown, then iterative MDS algorithm can be applied to compute relative coordinates of adjacent sensors.

The methods explained above can determine the relative positions of sensor nodes on the basis of their pairwise distances. These also require position alignment techniques to map the relative coordinates to physical coordinates based on three or more anchor sensors.

8.4.2 Ranging Estimation

Ranging estimation employs the widely used distance measurement model of RSSI. A circle centered in a sensor node bounds the maximal range for direct communication, which is called the "hop distance" of the sensor's radio signal. Nodes within one hop distance can directly communicate with each other, while nodes that are in more than one hop away relay messages hop by hop.

For example, there are four sensor nodes A, B, C, and D in Figure 8.7. The hop distance is $r_h.r_{ad}$ and is the distance between A and D, which can be induced with A's signal strength at the position of D.

8.4.3 Pairwise Distance Collection

Usually, a network of sensors is randomly and densely distributed. They are sufficiently connected and previous research indicates that the average connection degree of a node is between 5 and 15 in a general sensor network model. The essential operation in a pairwise distance collection is flooding by several selected sensor nodes. The procedure is described as follows: An anchor node is selected as the source sensor node to initialize a broadcast containing its ID, position, and hop count equal to 0. When another anchor node hears the broadcast, it uses the information in the received message to induce the average hop distance. The anchor node is then selected as a new source node and it initializes another broadcast later to collect more pairwise distances, as well as publish the average hop distance. Similarly, it can select some other nodes as source nodes to broadcast [48,49].

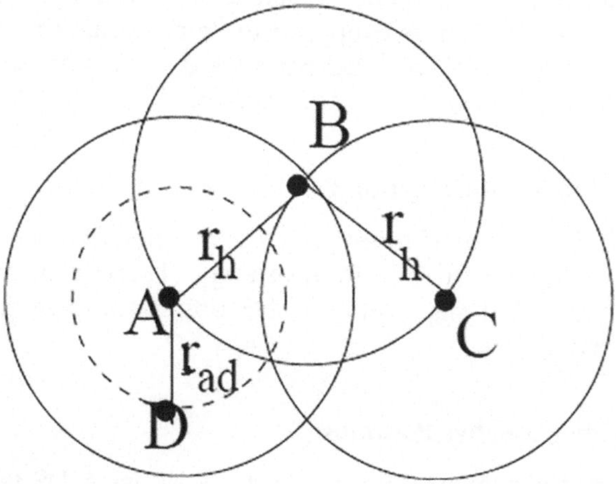

FIGURE 8.7
Signal strength and hop distance.

Consider the "n" number of sensors in a nertwork (sensor); there are a total of n (n – 1)/2 pairwise distances. The outcome of simulation results signifies that a source node broadcasts to the other nodes and typically collects 3–8% of all pairwise distances depending on the relative position of the source node in the network, connection degree of nodes, and hop distance. But, on a total, the number of pairwise distances collected don't rise linearly with the number of source nodes chosen, as there are several overlaps among the sets of broadcast routes, which decide the pairwise distance obtained by every source nodes' broadcast.

To reduce the total number of messages (or power consumption) sent or received by all nodes during the source nodes' broadcasts in the sensor network, we hope to initialize as few source sensors to broadcast and collect as many pairwise distances as possible. This requires that the broadcast from each source sensor can collect relatively more pairwise distances and that the overlap among sets of pairwise distances collected by every source node's broadcast is small.

8.4.4 Performance Study

To measure the performance of the algorithm with mean error, which is widely used in previous research works as in Equation 8.3:

$$error = \frac{\sum_{n-m+1}^{n} \left\| x_{est}^{i} - x_{real}^{i} \right\|_2}{(n-m)\times(radio-range)} \tag{8.3}$$

Here m and n are the number of sensors and the number of anchors, in that order. Low error indicates good performance of the method.

The source node selection is proposed to demonstrate the scheme. Here, we compare the number of collected pairwise distances based on the selection scheme and random source node selection.

8.4.5 Distributed Sensor Position Estimation

In this method, MDS is used to estimate the position of adjacent sensors in small regions which, in turn, form local maps. These local maps are then aligned to form a global map that describes position information of all sensors in the network.

8.4.5.1 Calculating Relative Positions

In the distributed sensor position method, the above MDS techniques are used in a distributed manner to estimate a local map for each group of adjacent sensors, and then these maps are aligned together based on the alignment method. The details of the distributed sensor position method are given in Chapter 8.

This method employs the widely used distance measurement model of RSSI. It is important to note that some other approaches for distance measure exist, like TOA, TDOA, AoA, and ultrasound, which can be applied here. They even generate more accurate distance measurements than RSSI, but they generally need complex hardware equipped in each sensor. In Chapter 8, it intends to use RSSI and simple hardware configuration to achieve competitive performance.

Based on the analysis of the challenges of the sensor position problem in real applications, the conditions that most existing sensor positioning methods fail to perform well are the anisotropic topology of the sensor networks and the complex terrain where the sensor networks are deployed [50]. In order to accurately position sensors in an anisotropic network and complex terrain, the distributed sensor position algorithm computes a series of local maps which are computed with multidimensional scaling. These local maps are then pieced together to get the approximation of the physical positions of the sensor nodes [51–53].

This method estimates the relative positions of sensor nodes based on their pairwise distances. We also need position alignment techniques to map the relative coordinates to physical coordinates based on three or more anchor sensors [54].

8.4.5.2 Aligning Relative Positions

Since we hope to compute the physical positions of sensors eventually, it is necessary to align the relative positions to physical positions with the aid of sensors with their known positions. For an adjacent group of sensors, at least three sensors' physical positions are needed in order to identify the physical positions of remaining nodes in the group in the 2D case. Thus, each group of adjacent sensors must contain at least three nodes with physical positions known, which may be anchors or nodes with physical positions calculated previously. The alignment usually includes shift, rotation, and reflection of coordinates.

8.4.5.3 Distribution of Nodes on Square Method

Since most applications depend on a successful position estimation (i.e., to compute their positions in some fixed coordinate system) it is of great importance to design efficient position estimation algorithms. In large-scale sensor networks, node localization can assist in routing [55]. The main advantage of node positioning in sensor networks is that it is an active field of research in wireless networking. Unfortunately, for a large number of sensor nodes, the straightforward solution of adding GPS to all nodes in the network is not feasible because:

- In the presence of dense forests, mountains, or other obstacles that block the line of sight from GPS satellites, GPS cannot be implemented.
- The power consumption of GPS will reduce the battery life of the sensor nodes and also reduce the effective lifetime of the entire network.
- In a network with large number of nodes, the production cost factor of GPS is an important issue.
- Sensor nodes are required to be small. But the size of the GPS and its antenna increases the sensor node form factor.

For these reasons, an alternate solution of GPS is required which is cost effective, rapidly deployable, and can operate in diverse environments.

MDS is used for sensor positioning in the square method, so to deploy sensor nodes in a large number, as shown in Figure 8.8. This method gives an accuracy that is 95% higher compared to a radio range of 85% achieved by square method [20,56].

In WSNs, sensors are purposely and accurately positioned at desirable locations in a controlled fashion following regular patterns [57]. Popular patterns including square, hexagon, and triangle that can be repeated to cover a continuous region without having any overlapping areas that are widely employed in practice because of its simplicity and ease of deployment.

The simulation results achieved for the triangulation and square method are as shown in Figure 8.9. When compared, it is observed that the

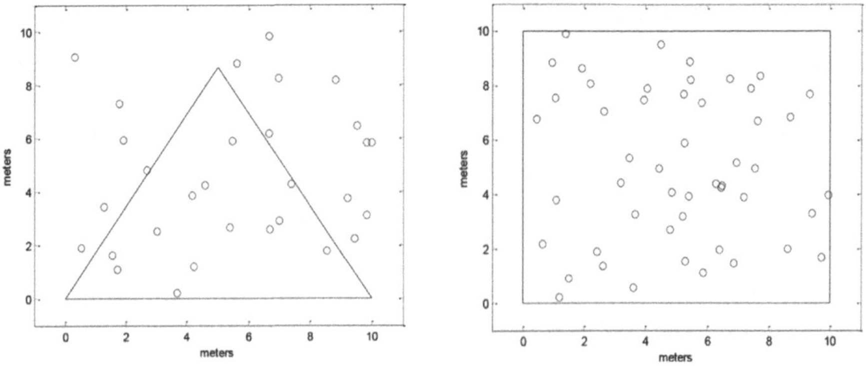

FIGURE 8.8
An example of randomly distributed sensors in triangulation method (left) and square method (right).

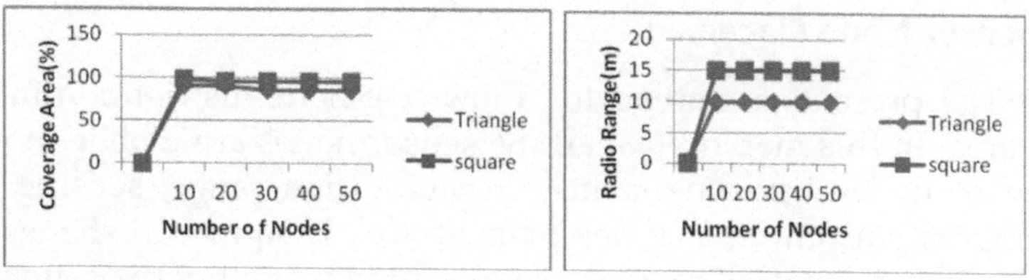

FIGURE 8.9
Comparison between existing triangulation and proposed square method using number of nodes vs. coverage area (left) and number of nodes vs. radio range (right).

distributionaof nodes on the square method outperforms triangulation by 9% in terms of coverage area and 33.33% in terms of radio range.

8.5 A Low-Cost Node Distribution Using the Hexagonal Method

Randomly deployed sensor nodes couldn't satisfy the requirements of wireless sensor networks in bad environmental conditions. Therefore, mobile sensor nodes are useful in a specified area for distribution of sensor nodes. This method is used to evaluate the accuracy of node distribution and deployment of sensor nodes to the appropriate positions of wireless sensor networks. Randomly deployed sensor nodes are depicted in Figure 8.10.

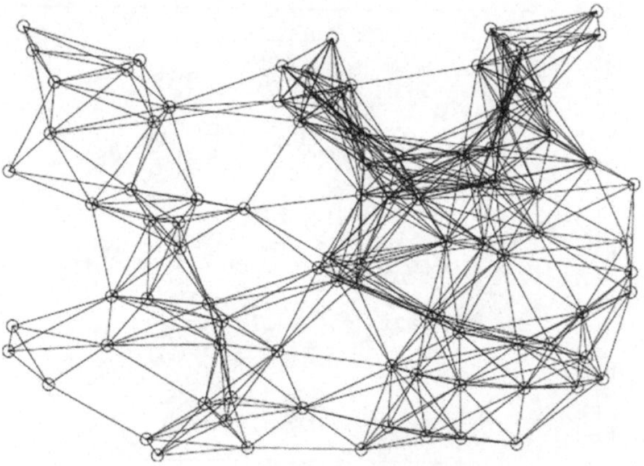

FIGURE 8.10
Randomly deployed sensor nodes.

8.5.1 Sensor Node Placement

Section 8.5.1 presents a method for a low-cost node distribution in a hexagonal area. In this method [58] all the sensor nodes are mobile in nature. The sensor nodes have the same communication range, sensing range, and computation power. The deployment area is square in shape, which is divided into number of regions are separated by half of the communication range. Further, the deployment area is divided into regular hexagons. MSNs will set themselves at the center of the hexagon on the instruction provided by the base station (BS) and the BS is located at one of the corners in the deployment area. The assumptions is that every sensor node falls within the deployment area when deployed randomly and the total number of sensor nodes deployed is greater than or equal to the total number of desired locations in the deployment area. Random deployment requires a larger number of sensor nodes to achieve the required level of coverage which increases the system cost. So, for optimum and efficient utilization of resources and for better control and management, the RSPBFA algorithm is preferred.

The simulation results that are obtained are depicted in Figure 8.11. The 1000 nodes are deployed in a hexagonal method with a square region. In the hexagon C (0, 0), C represents center, the length of the side is 20 m, and the area of the hexagon is shown in Equation 8.4.

$$A = (3\sqrt{3}/2)S^2 \tag{8.4}$$

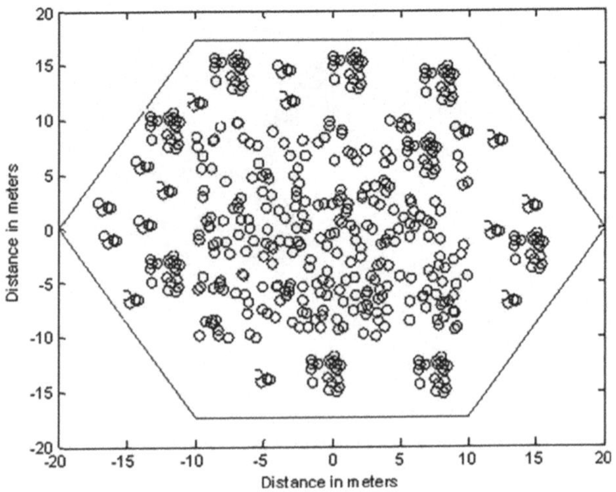

FIGURE 8.11
1000 nodes deployed randomly on a hexagonal area.

8.5.2 Challenges of the Hexagonal Method

As mentioned in the preceding Section 8.5.1, most of the sensor positioning methods of research that exist try to offer accurate sensor positioning for a network of sensors. Wireless sensor networks have limited energy availability, while sensor positioning usually involves energy-consuming computation and communication. It is always desirable to reduce the energy costs for sensor positioning. Among the many approaches to reduce energy caused by sensor positioning, one of the most effective methods is to reduce computational cost and communication costs for sensor positioning. So, it is necessary to develop sensor positioning methods that are able to locate sensors only on the hexagonal method for sensor position estimation in wireless sensor networks. It is also of note that many applications and operations in sensor networks only require positioning the information of a few sensors. In this method, the sensors are deployed without any ranging error, a low computational cost, and a high coverage area. Figure 8.12 shows the comparison of the existing methods with the proposed method.

On comparing the performance of the three methods (triangular, square, and hexagonal) the hexagonal method provides a much better coverage area.

8.6 2D Iterative Routing Algorithm

The efficiency of a 2D iterative routing process is evaluated as per the properties of perceptual transparency, the shortest path, robustness, cost of computation, radio range, range error, and the numer of nodes. In Section 8.6, a new method is proposed, the iterative routing method, which is used in order

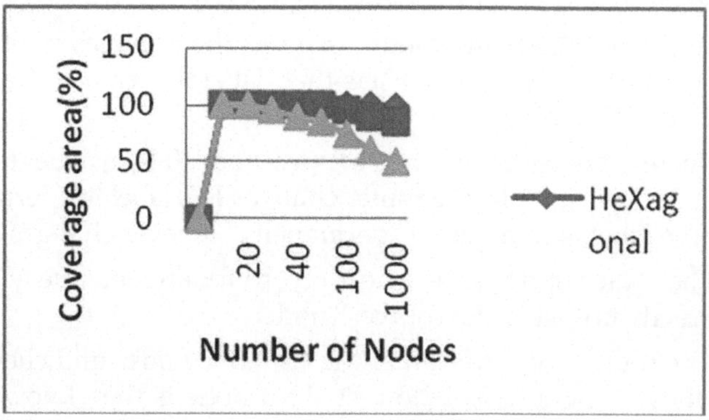

FIGURE 8.12
Number of nodes vs. coverage area in a WSN following existing triangular, square and proposed hexagonal patterns.

to realize most of these requirements. The iterative routing method can be chosen for decreasing computational cost [59,60].

Kulkarni et al. [61] proposed a bio-inspired algorithm for determining the coordinates of nodes in a WSN in a distributed and iterative fashion. Jasper [62] made some improvements [61] to the issues of node deployment, position estimation, energy-aware clustering, and data-aggregation. The accuracy of [62] is better when compared to [61]. Gupta et al. [63] proposed decentralized and adaptive routing strategies for position estimation and measurement of distances. Two methods are required in order to estimate node position and finding out the shortest path measurement. The proposed method does not need a separate method to find the shortest path measurement.

8.6.1 IR Algorithm

The router develops a graph of the network. Then it identifies source and destination nodes, for example R1 and R2. The router builds then a matrix, called the "adjacency matrix." In the adjacent matrix, a coordinate indicates weight (i, j). Suppose the weight of a link between nodes R_i and R_j. If there is no direct link between R_i and R_j, this weight is identified as "infinity."

8.6.1.1 Procedure for the 2D IR Algorithm

Step 1: The router builds a status record for each node on the network. The record contains the three following fields:

- Predecessor field: Shows the previous node
- Length field: Shows the sum of the weights from the source to that node
- Label field: Shows the status of the node; each node has one status mode, either "permanent" or "tentative."

Step 2: In the next step, the router assumes the parameters of the status record (for all nodes) and sets their label to "tentative" and their length to "∞".

Step 3: During this step, the router sets a T-node. If R_1 is to be the source of T-node, for example, the router changes R_1's label to "permanent." Once the label is changed to "permanent," it never changes again.

Step 4: The router updates the status record for all tentative nodes that are directly linked to the source T-node.

Step 5: The router goes over all of the tentative nodes and chooses the one whose weight to R_1 is lowest. That node is then known as the destination T-node.

Step 6: If the new T-node is not R_2 (the intended destination), the router goes back to step 5.

Step 7: If this node is R_2, the router extracts its previous node from the status record and does this until it arrives at R_1. This list of nodes shows the best route from R_1 to R_2, as shown in Figure 8.13. The IR Algorithm is compared with ant colony optimization algorithm, as shown in Figure 8.14.

The hexagonal distribution method 3 proposed higher accuracy because more number of sensor nodes are placed to find position estimation and shortest path distance [64–67]. The total cost of computation for this technique is 170.5421. While method [61] costs around 261.5371 and 388.4575 is achieved by method [63]. The proposed method 3 has a range error which accounts to 1% and 8% is achieved by method [61] and 9% is obtained by method [63].

The iterative routing algorithm (IRA) is better when compared to the existing particle swarm optimization (PSO) and ant colony optimization (ACO),

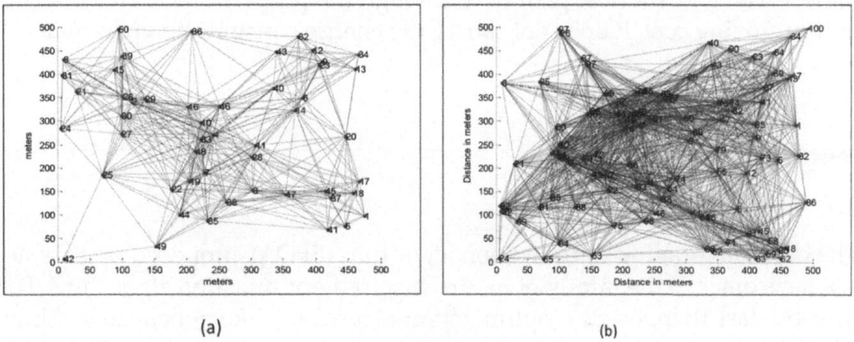

FIGURE 8.13
(a) 50 nodes randomly generated to cover the area 500 × 500 m² and (b) 100 nodes to cover 500 × 500 m² area using IRA and find out shortest path from source and destination nodes.

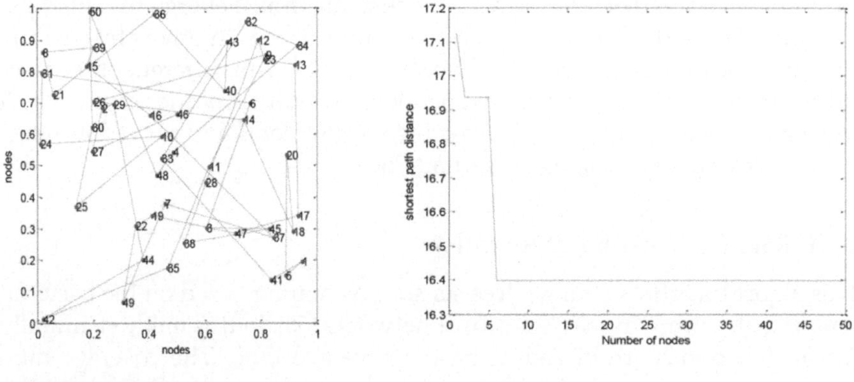

FIGURE 8.14
(Left) Positions estimated by existing ant colony optimization (right) shortest path using existing ant colony optimization. (From Anuj K. Gupta et al., *International Journal of Modeling and Optimization*, Vol. 2, No. 1, February 2012.)

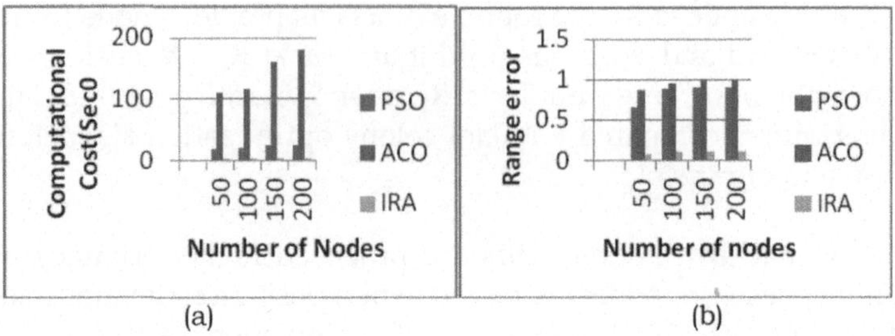

FIGURE 8.15
(a) Nodes vs. computational cost and (b) nodes vs. range error for proposed IRA with existing PSO and ACO.

as shown in Figure 8.15. In this method it can find the computational cost, range error, and shortest path only. Though the proposed IRA outperforms in minimizing cost, it does not satisfy the energy consumption parameter.

8.7 RSPBFA Protocol

The bacteria foraging optimization algorithm (BFOA), proposed by Passino, is a newcomer to the family of nature-inspired optimization algorithms. For over the last five decades, optimization algorithms like genetic algorithms (GAs), evolutionary programming (EP), evolutionary strategies (ES), which draw their inspiration from evolution and natural genetics, have dominated the realm of optimization algorithms. Recently, natural swarm-inspired algorithms like PSO and ACO have found their way into this domain and proved their effectiveness. Bacteria search is done to maximize energy obtained per unit time. Individual bacterium also communicates with others by sending signals. Researchers are trying to hybridize BFOA with different other algorithms in order to explore its local and global search properties separately. It has already been applied to many real-world problems and proved its effectiveness over many variants of GA and PSO.

8.7.1 RSPBFA Algorithm Description

This algorithm offers a range–free sensor positioning based on the bacterial foraging algorithm in wireless sensor networks [68]. In this method, initially the anchor nodes are placed using the coverage ratio. The coverage ratio depends on the network size. Then the anchor nodes use the BFO algorithm to estimate the distance between the unknown sensor nodes using neighbor density.

8.7.2 Estimation of Coverage Ratio

Every sensor knows its own position. It communicates with the mobile nodes and other sensor nodes can change their positions by using the other nodes' information. The coverage ratio (β) of the sensor network is evaluated using the following Equation 8.5:

$$\beta = \frac{CS_i}{Z}, i \varepsilon X \tag{8.5}$$

where CS_i = coverage of a sensor i
X = set of the nodes
Z = total size (based on the network space considered)

The sequence of steps employed in implementing the proposed RSPBFA-based method is presented in Figure 8.16.

Anchor node deployment in WSNs is application dependent and affects the performance of the routing protocol. The deployment can be either deterministic or randomized. In deterministic deployment, the sensors are manually placed and data is routed through fixed paths; but in random node

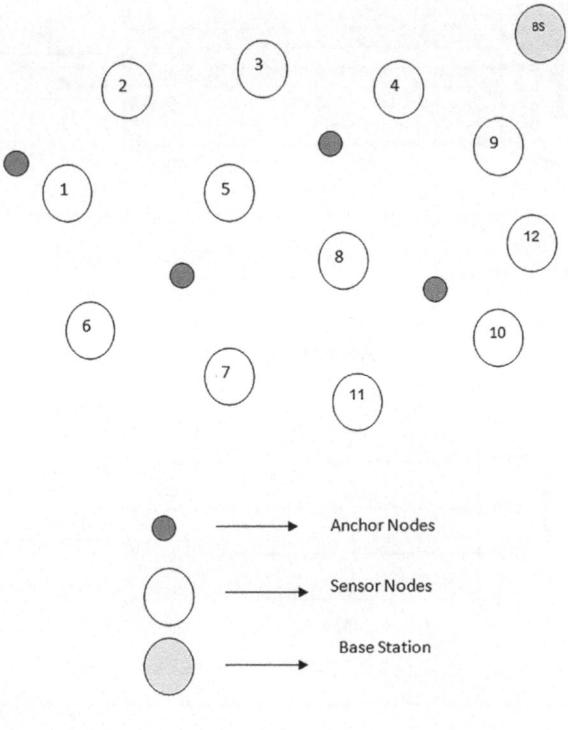

FIGURE 8.16
Deployment of anchor nodes.

deployment, the sensor nodes are scattered randomly creating an infrastructure in an ad hoc manner. Hence, random deployment raises several issues as coverage, optimal clustering, etc. which need to be addressed.

Figures 8.17 to 8.21 show the results of delay, delivery ratio, packet drop, residual energy, and energy consumption by varying the number of nodes as 50, 100, 150, and 200 for the CBR traffic in RSPBFA and ABC protocols. On

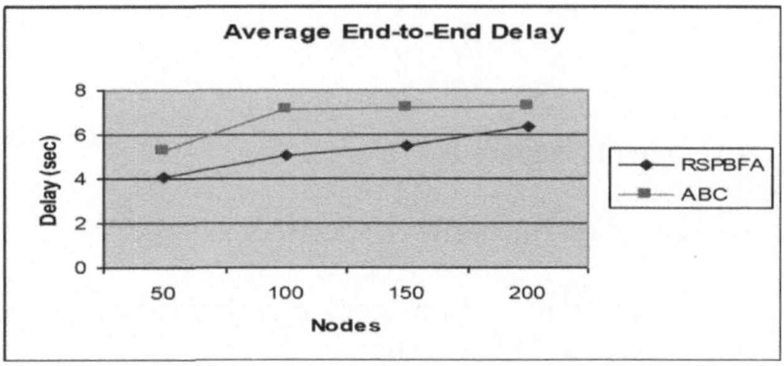

FIGURE 8.17
Nodes vs. delay (Sec).

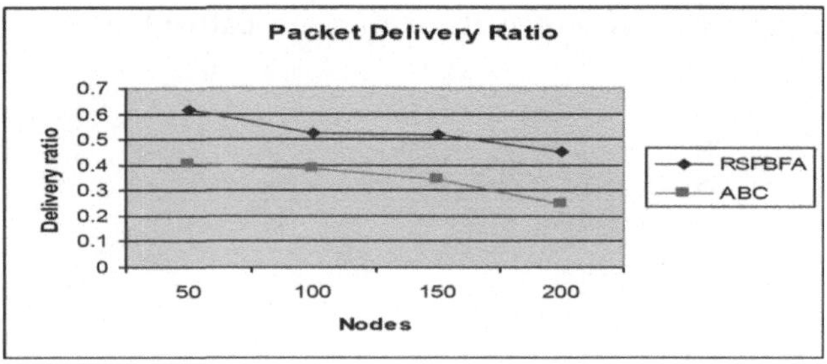

FIGURE 8.18
Nodes vs. packet delivery ratio.

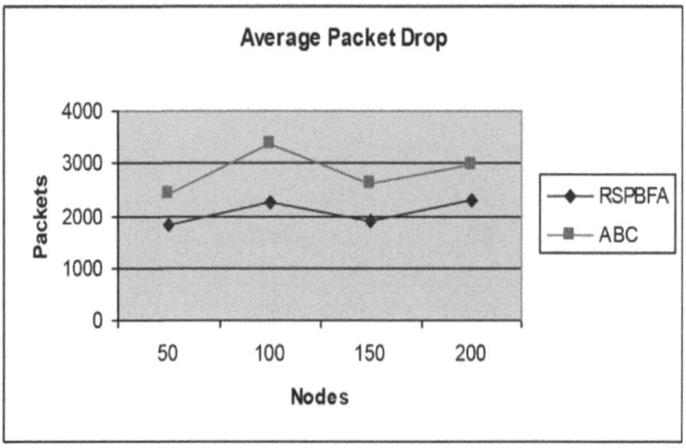

FIGURE 8.19
Nodes vs. packet drop.

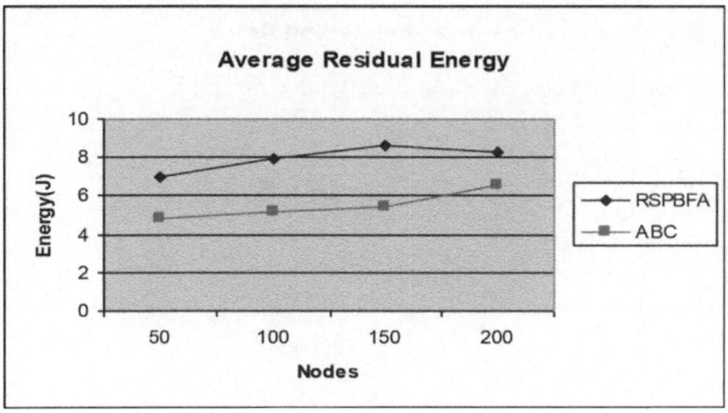

FIGURE 8.20
Nodes vs. residual energy.

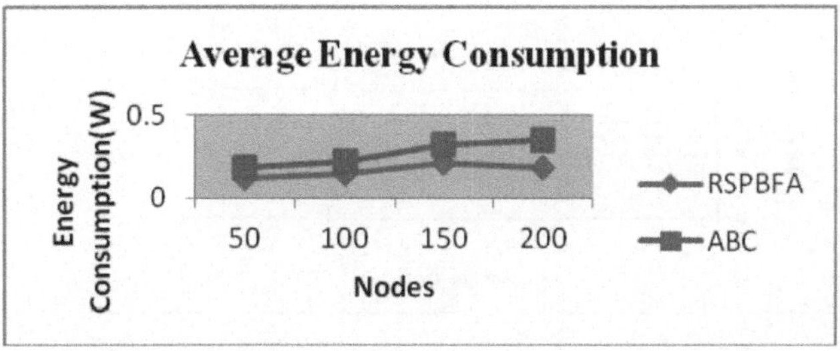

FIGURE 8.21
Nodes vs. energy consumption.

comparing the two protocol performances, the RSPBFA outperforms existing ABC by 21% in terms of delay, 48% in terms of delivery ratio, 22% in terms of packet drop, 31% in terms of residual energy, and 37% in terms of energy consumption.

8.7.3 Based on Range

The range of transmission is varied as 250, 300, 350, and 400 m/s for 200 nodes. Figure 8.22 depicts the comparison of range vs. average end-to-end delay of proposed RSPBFA with existing ABC [12] algorithm.

Figures 8.22 to 8.26 show the results of delay, delivery ratio, packet drop, residual energy, and energy consumption by varying the range as 250, 300, 350, and 400 for the CBR traffic in RSPBFA and ABC protocols. On comparing the performance of the two protocols, the proposed RSPBFA outperforms existing ABC by 46% in terms of delivery ratio, 15% in terms of delay, 11% in terms of packet drop, 49% in terms of energy consumption, and 25% in terms of residual energy.

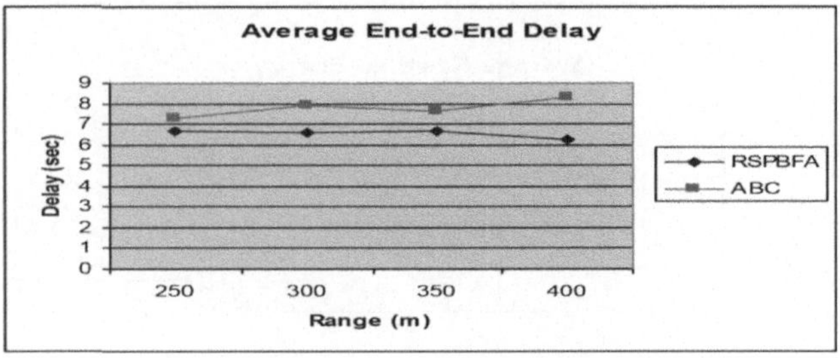

FIGURE 8.22
Range vs. delay.

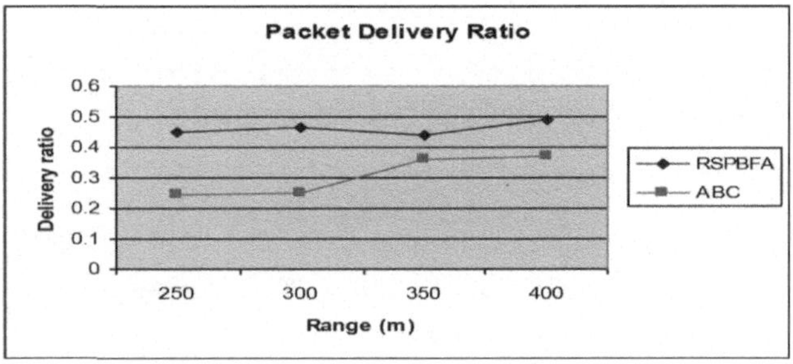

FIGURE 8.23
Range vs. packet delivery ratio.

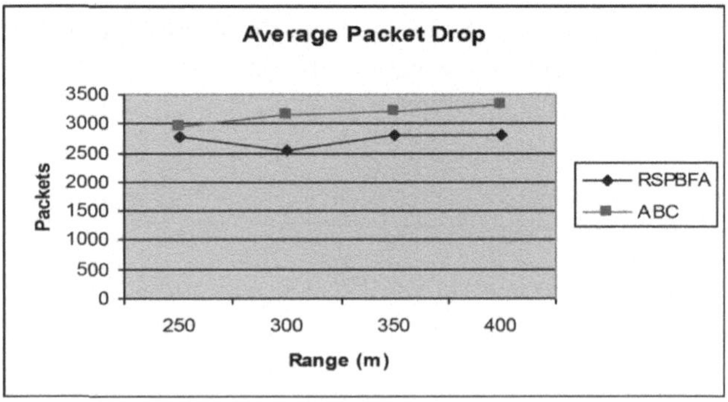

FIGURE 8.24
Range vs. packet drop.

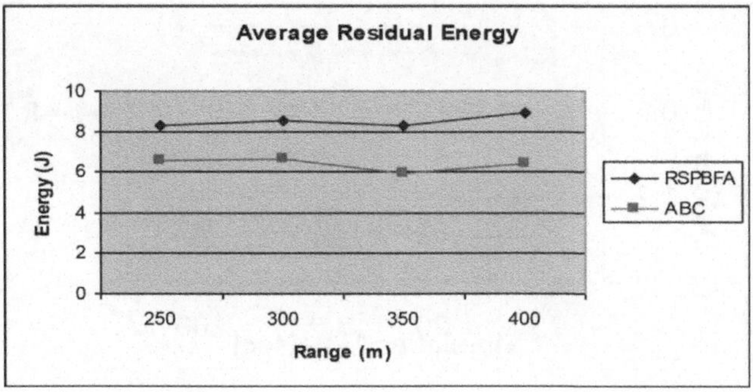

FIGURE 8.25
Range vs. residual energy (J).

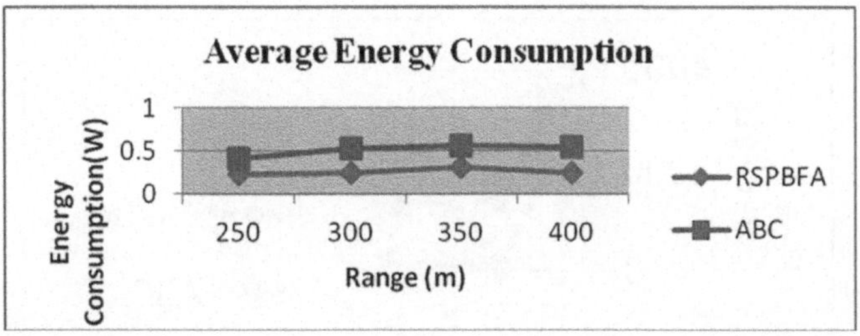

FIGURE 8.26
Range vs. energy consumption (W).

8.7.4 Based on Simulation Time

The simulation time changes from 25 to 100 seconds for 100 nodes.

Figures 8.27 to 8.31 show the results of delay, packet drop, delivery ratio, and residual energy by varying the simulation time from 25 to 100 seconds for RSPBFA and ABC protocols. When comparing the performance of the

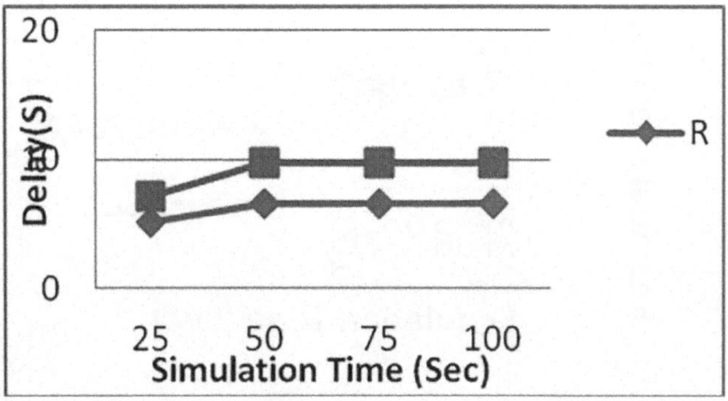

FIGURE 8.27
Simulation time vs. delay.

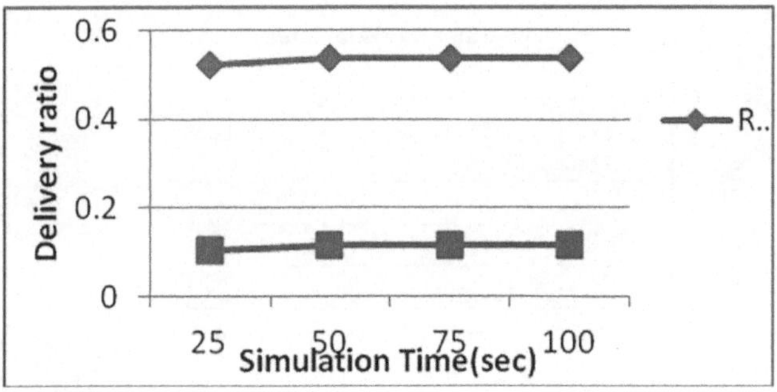

FIGURE 8.28
Simulation time vs. packet delivery ratio.

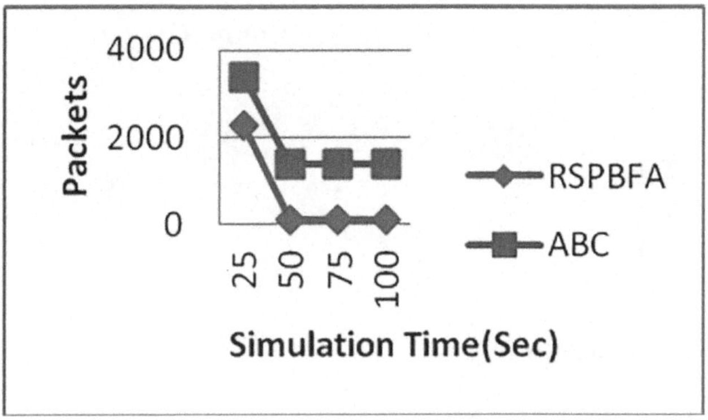

FIGURE 8.29
Simulation time vs. packet drop.

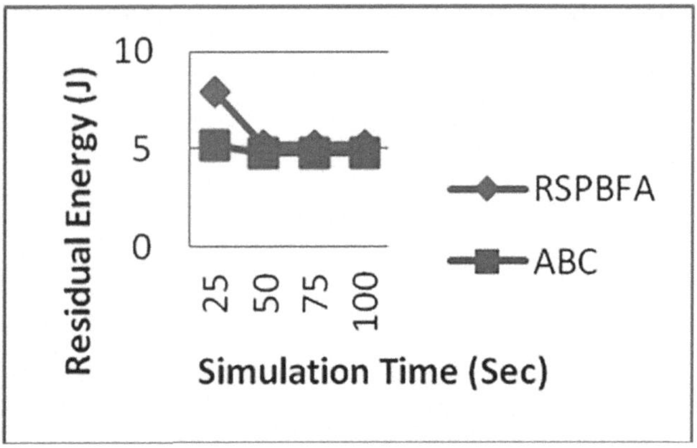

FIGURE 8.30
Simulation time vs. residual energy.

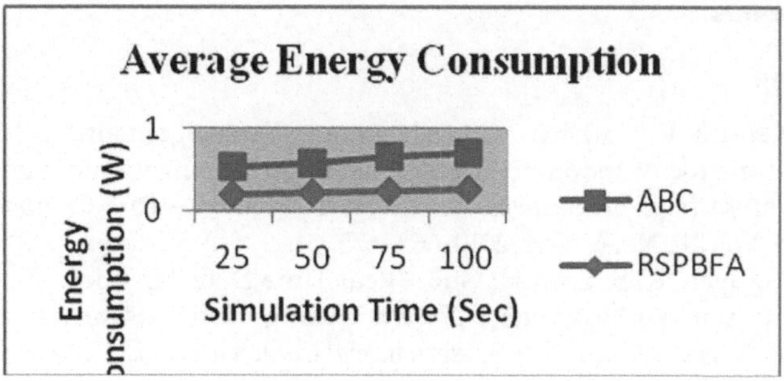

FIGURE 8.31
Simulation time vs. energy consumption.

two protocols, the proposed RSPBFA outperforms existing ABC by 35% in terms of delay, 78% in terms of delivery ratio, 76% in terms of packet drop, 16% in terms of residual energy, and 42% in terms of energy consumption.

8.8 Conclusions and Future Work

The primary functions of sensor networks include sensor positioning, tracking, navigation, and sensing. Sensor positioning in sensor networks is a fundamental and crucial issue, because sensor position information is a prerequisite for positioning and tracking objects, targeting, routing in network, and many other tasks. However, in most cases, sensors are randomly deployed in inaccessible terrain. It is desirable to develop robust and efficient algorithms that enable sensors to perform sensor positioning and distance measurements. In Chapter 8, four new techniques for sensor positioning and distance measurements have been discussed. The first three algorithms were implemented and tested in MATLAB® 2011b under a WindowsXP operating system and the fourth algorithm was implemented and tested in NS-2 simulator.

The observations made from this study signifies that the distribution of nodes on the hexagonal method and square method are powerful and resourceful techniques in wireless sensor networks, which are very useful in position estimation with reduced expense and greater coverage area. This method was tested with 2D scaling and therefore 3D scaling can be developed. The 2D iterative routing algorithm can be used to perform the operation of distance measurements, and it is one of the best among existing positioning algorithms. This is designed for 2D techniques and, as such, algorithms can be developed for multidimensional techniques. These proposed methods are developed for 2D scaling techniques for sensor positioning and distance measurements in wireless sensor networks.

References

1. Raghavendra V. Kulkarni, Ganesh Kumar Venayagamoorthy. "Bio-inspired Algorithms for Autonomous Deployment and Localization of Sensor Nodes." *IEEE Transactions on Systems, Man, and Cybernetics—Part C: Applications and Reviews* Vol. 40, No. 6, Nov. 2010, 663–675.
2. H. A. Nguyen, K. S. Low, H. Guo. "Real Time Determination of Sensor Node Location in a Wireless Sensor Network using Particle Swarm Optimization." *Proceedings of the 10th WSEAS International Conference on Evolutionary Computing* 2009, 140–145.
3. Hyungmin Park, Ji-Hyeong Han, Jong-Hwan Kim. "Swarm Intelligence-based Sensor Network Deployment Strategy." *IEEE World Congress on Computational Intelligence* July 2010, 4210–4215.
4. Wu Xiaoling, Shu Lei, Wang Jin, Jinsung Cho, Sungyoung Lee. "Energy-efficient Deployment of Mobile Sensor Networks by PSO." *AP Web Workshops* 2006, 1–10.
5. Amjad Osmani. "Design and evaluation of two distributed methods for sensors placement in Wireless Sensor Networks." *Journal of Advances in Computer Research* 2011, 13–26.
6. Nikitha Kukunuru, Babu Rao Thella, Rajya Lakshmi Davuluri. "Sensor Deployment Using Particle Swarm Optimization." *International Journal of Engineering Science and Technology* Vol. 2, No. 10, 2010, 5395–5401.
7. Wen-Hwa Liao, Yucheng Kao, Ying-Shan Li. "A sensor deployment approach using glowworm swarm optimization algorithm in wireless sensor networks." *Expert Systems with Applications* 2011, 12180–12188.
8. Nadjib Aitsaadi, Nadjib Achir, Khaled Boussetta, Guy Pujolle. "Multi-Objective WSN Deployment: Quality of Monitoring, Connectivity and Lifetime." *IEEE ICC Proceedings* 2010, 1–6.
9. Roghayeh Soleimanzadeh, Bahareh J. Farahani, Mahmood Fathy. "PSO based Deployment Algorithms in Hybrid Sensor Networks." *IJCSNS International Journal of Computer Science and Network Security*, Vol. 10, No. 7, July 2010, 167–171.
10. Dey, N., Ashour, A. S., Shi, F., Fong, S. J., Sherratt, R. S. "Developing residential wireless sensor networks for ECG healthcare monitoring." *IEEE Transactions on Consumer Electronics* Vol. 63, No. 4, 2017, 442–449.
11. Ashour, A. S., Shi, F., Sherratt, R. S. "Wireless capsule gastrointestinal endoscopy: Direction of arrival estimation based localization survey." *IEEE Reviews in Biomedical Engineering Methodological Reviews.*
12. Celal Ozturk, Dervis Karaboga, Beyza Gorkemi. "Probabilistic Dynamic Deployment of Wireless Sensor Networks by Artificial Bee Colony Algorithm." *Sensors* 11, 2011, 6056–6065.
13. G. Mao, B. Fidan, B. D. O. Anderson. "Wireless sensor network localization techniques." *Comput. Netw.* Vol. 51, No. 10, July 2007, 2529–2553.
14. Quan LIU, Ping Ren, Zude Zhou, "Three-dimensional Accurate Positioning Algorithm based on Wireless Sensor Networks." *Journal of Computers* Vol. 6, No. 12, December 2011, 2582–2589.
15. Naveed Salman, Mounir Ghogho, A. H. Kemp. "Optimized Low Complexity Sensor Node Positioning in Wireless Sensor Networks." *IEEE* 2013.
16. C. Savarese. "Robust positioning algorithms for distributed ad hoc wireless sensor networks." Master's thesis, University of California at Berkeley, 2002, 1–10.

17. Junho Park, Hyuk Park, Dong-ook Seong, Jaesoo Yoo. "A Sensor Positioning Scheme with High Accuracy in Nonuniform Wireless Sensor Networks." *International Journal of Distributed Sensor Networks* Vol. 2013, Article ID 507605.
18. Junho Park, Hyuk Park, Dong-ook Seong, Jaesoo Yoo, "A Sensor Positioning Scheme with High Accuracy in Nonuniform Wireless Sensor Networks." *International Journal of Distributed Sensor Networks* Vol. 2013, Article ID 507605.
19. Chunlai Chai. "Efficient Intelligent Localization Scheme for Distributed Wireless Sensor Networks." *Journal of Computers*, Vol. 4, No. 11, Nov. 2009, 1159–1165.
20. Tian He, Chengdu Huang, Brian M. Blum, John A. Stankovic, Tarek Abdelzaher. "Range-Free Localization Schemes for Large Scale Sensor Networks" MobiCom '03, September 14–19, 2003, San Diego, California, 1–15.
21. Yun Wang, Yanping Zhang, Jiangbo Liu, Rahul Bhandari. "Coverage, Connectivity, and Deployment in Wireless Sensor Networks." S. Patnaik et al. (eds.) *Recent Development in Wireless Sensor and Ad-hoc Networks, Signals and Communication Technology* © Springer India 2015, DOI 10.1007/978-81-322-2129-6_2.
22. A. Howard, M. Mataric, G. S. Sukhatme. "Localization for mobile robot teams using maximum likelihood estimation." *Proceedings of the IEEE/RSJ International Conference on Intelligent Robots and Systems*, Lausanne, Switzerland, October 2002, 434–459.
23. S. Roumeliotis, G. Bekey. "Collective localization: A distributed kalman filter approach." *Proceedings of IEEE International Conference on Robotics and Automation* (ICRA), San Francisco, May 1999, 2958–2965.
24. D. Fox, W. Burgard, H. Kruppa, S. Thrun. "A probabilistic approach to collaborative multi-robot localization." *Autonomous Robots* Vol. 8, No. 3, 2000, 325–344.
25. A. Savvides, C.-C. Han, M. Srivastava. "Dynamic fine-grained localization in Ad-hoc networks of sensors." The 7th ACM Int. Conf. on Mobile Computing and Networking (Mobicom), 166–179, Rome, Italy, July 2001.
26. X. Cheng, A. Thaeler, G. Xue, D. Chen. "Tps: A time-based positioning scheme for outdoor wireless sensor networks." *Proceedings of the 23rd Conference of the IEEE Communications Society* (IEEE INFOCOM), Hong Kong, China, March 2004.
27. N. B. Priyantha, A. Chakraborty, H. Balakrishnan. "The cricket location support system." *Proc. 6th International Conference on Mobile Computing and Networking* (ACM Mobicom), 2000.
28. S. Tabbane. "Location management methods for third-generation mobile systems." *IEEE Communications* Vol. 35, 1997, 72–85.
29. N. Bulusu. "Self-Configuring Localization Systems." PhD thesis, University of California, Los Angeles, 2002.
30. S. Ratnasamy, M. Handley, R. Karp, S. Shenker. "Topologically-aware overlay construction and server selection." *Proc. IEEE Infocom*, 1190–1199, New York, NY, 2002.
31. F. Ye, H. Luo, J. Cheng, S. Lu, L. Zhang. "A two-tier data dissemination model for large-scale wireless sensor networks." *Proceedings of Eighth Annual International Conference on Mobile Computing and Networking* (Mobicom), Atlanta, GA, September 2002, 148–159.
32. S. Thrun, D. Fox, W. Burgard, F. Dellaert. "Robust Monte Carlo localization for mobile robots." *Artificial Intelligence* Vol. 128, No. 1, Summer 2001, 99–141.

33. A. Savvides, W. Garber, S. Adlakha, R. Moses, M. Srivastava. "On the error characteristics of multihop node localization in wireless sensor networks." *Proceedings of Second International Workshop on Information Processing in Sensor Networks* (IPSN 2003), Palo Alto, CA, April 2003, 317–332.
34. D. Niculescu, B. Nath. "Error characteristics of ad hoc positioning systems", *Proceedings of Fifth International Symposium on Mobile Ad Hoc Networking and Computing* (Mobihoc 2004), Tokyo, Japan, May 2004.
35. K. Chintalapudi, R. Govindan, G. Sukhatme, A. Dhariwal, "Ad-hoc localization using ranging and sectoring." *Proceedings of the 23rd Conference of the IEEE Communications Society* (IEEE INFOCOM), Hong Kong, China, March 2004.
36. T. Eren, D. Goldenberg, W. Whiteley, Y. Yang, A. Morse, B. Anderson, P. Belhumeur. "Rigidity, computation, and randomization in network localization." *Proceedings of the 23rd Conference of the IEEE Communications Society* (IEEEINFOCOM), Hong Kong, China, March 2004.
37. J. Aspnes, D. Goldenberg, Y. R. Yang. "On the computational complexity of sensor network localization." *Proceedings of First International Workshop on Algorithmic Aspects of Wireless Sensor Networks* (ALGOSENSORS 2004), Turku, Finland, July 2004.
38. Hua Chen, Min Luo, Baolin Sun. "Energy Efficient Based Maximal Lifetime Routing in Wireless Sensor Networks." *International Journal of Future Generation Communication and Networking* Vol. 7, No. 3, 2014.
39. Gyöngyi Kocsisné Szilágyi, Attila Kocsis. "A special localization algorithm in Wireless sensor networks for telemetry application." *Annales Mathematicae et Informaticae* Vol. 44, 2015, 121–130.
40. M. B. Nirmala. "Enhanced Voting based Secure Localization for Wireless Sensor Networks." *I. J. Computer Network and Information Security* Vol. 12, 2015, 52–59, Published Online November 2015 in MECS (http://www.mecs-press.org/)DOI: 10.5815/ijcnis.2015.12.06.
41. Binh, H. T. T., Hanh, N. T., and Dey, N. "Improved cuckoo search and chaotic flower pollination optimization algorithm for maximizing area coverage in wireless sensor networks." *Neural Computing and Applications*, 2016, 1–13.
42. T.S. Rappaport. *Wireless Communications: Principles and Practice.* Prentice Hall PTR, 1996.
43. D. Niculescu, B. Nath. "Ad hoc positioning system (aps) using aoa." *Proceedings of the 20th Annual Joint Conference of the IEEE Computer and Communications Societies* (INFOCOM-20), April 2003.
44. J. Beutel. "Geolocation in a picoradio environment." Master's thesis, ETH Zurich, 1999.
45. Xiang Ji. "Localization Algorithms For Wireless Sensor Network Systems." Thesis, The Pennsylvania State University, The Graduate School, Department of Computer Science and Engineering, 2004
46. P.E. Green, F.J.Caromone, S.M. Smith. *Multidimensional Scaling: Concepts and Applications.* Newton, MA: Allyn and Bacon, 1989.
47. A. Morrison, G. Ross, M. Chalmers. "Combining and comparing clustering and layout algorithms." Technical report, Computing Science Department, University of Glasgow, 2003.

48. N. Pushpalatha, Dr. B. Anuradha. "Study of Various Methods of Wireless Ad-hoc Sensor Networks using Multidimensional Scaling for Position Estimation." *Global Journal Engineering and Applied Sciences* ISSN2249-2631, 2249–2623 GJEAS Vol. 1, No. 3, 2011, 76–81.
49. N. Pushpalatha, Dr. B. Anuradha. "Distribution of Nodes on Square Method for Wireless Sensor Networks." *International Journal of Computer Science and Telecommunications* Vol. 3, Issue 1, January 2012, 74–77.
50. Xiangi Ji, Hongyuan Zha. "Sensor positioning in wireless Adhoc sensor Networks using multidimensional Scaling." *IEEE IFOCOM* 2004, 2652–2661.
51. N. Bulusu, J. Heidemann, D. Estrin. "Adaptive beacon placement." *Proceedings of the 21st International Conference on Distributed Computing Systems* (ICDCS-21), Phoenix, Arizona, April 2001, 489–498.
52. N. Bulusu, J. Heidemann, D. Estrin. "Gps-less lowcost outdoor localization for very small devices." *IEEE Personal Communications* Vol. 7, No. 5, October 2000, 28–34.
53. A. Ward, A. Jones, A. Hopper. "A new location technique for the Active Office." *IEEE Personal Communications Magazine* Vol. 4, No. 5, 1997.
54. S. Roumeliotis, G. Bekey. "Collective localization: A distributed kalman filter approach." *Proceedings of IEEE International Conference on Robotics and Automation* (ICRA), San Francisco, May 1999, 2958–2965.
55. M. Mauve, J. Widmer, H. Hartenstein. "A Survey on Position Based Routing in Mobile Ad-hoc Networks." *IEEE Network Magazine,* Vol. 15, No. 6, November 2001, 30–39.
56. L. Schwiebert, S. K. S. Gupta, J. Weinmann. "Research challenges in wireless networks of biomedical sensors." *Mobile Computing and Networking* 2001, 151–165.
57. Borg, I., P. Groenen. *Modern Multidimensional Scaling Theory and Applications.* Springer, 1997.
58. N. Pushpalatha, Dr. B. Anuradha. "A Low Cost Nodes Distribution on Hexagonal Method for Wireless Sensor Networks." *International Journal of Engineering Research and Applications* (IJERA) ISSN: 2248–9622 www.ijera.com. Vol. 2, Issue 4, July–August 2012, 1718–1721.
59. N. Pushpalatha, Dr. B. Anuradha. "A Two-Dimensional IR Algorithm for Position Estimation in Wireless Sensor Networks." *International Journal of Computer Science and Technology,* ISSN: 0976-8491(Online) | ISSN: 2229-4333, IJCST Vol. 3, Issue 2, April–June 2012, 35–39.
60. N. Pushpalatha, Dr. B. Anuradha. "A Comparative Analysis of WSN Sensor Positioning Method using Iterative Routing Algorithm with Conventional Methods." *International Journal of Computer Applications* (0975–8887) Vol. 53, No. 7, September 2012, 31–35.
61. Raghavendra V. Kulkarni, Ganesh K. Venayagamoorthy, Maggie X. Cheng. "Bio-Inspired Node Localization in Wireless Sensor Networks." *Proceedings of the 2009 IEEE International Conference on Systems, Man, and Cybernetics.* San Antonio, TX, October 2009, 205–210.
62. Mr. J. Jasper Gnana Chandran. "An Energy Efficient Localization Technique Using Particle Swarm Optimization in Mobile Wireless Sensor Networks." *American Journal of Scientific Research* ISSN 1450-223X Issue 8, 2010, 33–48 *Euro Journals Publishing, Inc.* 2010 http://www.eurojournals.com/ajsr.htm.

63. Anuj K. Gupta, Harsh Sadawarti, Anil K. Verma. "MANET Routing Protocols Based on Ant Colony Optimization." *International Journal of Modeling and Optimization* Vol. 2, No. 1, February 2012.
64. Al-Karaki, Jamal N., Kamal, Ahmed E. "Routing techniques in wireless sensor networks: A survey." *IEEE Wireless Commun.* Vol. 11, No. 6, 2004, 6–28.
65. Liu, B., Towsley, D. "A study of the coverage of large-scale sensor networks." IEEE International Conference on Mobile Ad-Hoc and Sensor Systems, 475–483. IEEE 2004.
66. Liu, B., Brass, P., Dousse, O., Nain, P., Towsley, D. "Mobility improves coverage of sensor networks." *Proceedings of the 6th ACM International Symposium on Mobile Ad Hoc Networking and Computing*, ACM (2005), 300–308.
67. N. Pushpalatha, Dr. B. Anuradha. "Shortest path position estimation between source and destination nodes in wireless sensor networks with low cost." *International Journal of Emerging Technology and Advanced Engineering* ISSN 2250–2459, Vol. 2, Issue 4, April 2012, 6–12.
68. N. Pushpalatha, Dr. B. Anuradha. "Range-free Sensor Positioning Based on Bacterial Foraging Algorithm (BFO) in Wireless Sensor Networks." *International Journal of Computer Applications* (0975–8887) Vol. 122, No. 9, July 2015, 35–40. Doi >10.5120/21731-4901.

9

Internet of Things in Healthcare Wearable and Implantable Body Sensor Network (WIBSNs)

Anu Rathee, T. Poongodi, Monika Yadav, and Balamurugan Balusamy

CONTENTS

9.1 Introduction

In the beginning of 21st century, no one could have predicted the huge impact of the Internet and the IoT in our daily lives. Nowadays, IoT is recognized as the new revolution of the Internet and the practice of IoT has been improved drastically in various circumstances. IoT links different objects to the Internet, which enable data and insights that have never existed before. An IoT device is a computing device that connects objects wirelessly to a network and provides the ability to transmit data; these are viewed as the things in the Internet of Things. A huge number of entities are physically interconnected with the Internet in IoT–based infrastructure which enables easy connectivity and effective communications. The embedded technology in the objects assists them to communicate internally as well as externally, which in turn helps to make further decisions. The "things" in the IoT [1–3] could be a human with a heart monitor or an automobile with built-in sensors, i.e., objects that have been allotted an IP address. Those objects are capable of collecting and transferring data over a network without any human intervention or assistance. A pacemaker is a device which can be fitted in the chest or abdomen and it is used to manage the heart with some abnormal rhythms. The functioning of Internet of Things is inevitable in many areas such as agriculture, industry, education, smart cities, healthcare, etc.

The IoT offers appropriate solution for various applications including the industrial sector, traffic congestion, security, emergency services, waste management, logistics, smart cities, and healthcare, as shown in Figure 9.1. The IoT concept in the real world has become possible via the integration of different enabling technologies including:

- Identification, sensing, and communication technologies
- Middleware architecture in the IoT follows the service oriented architecture (SOA) approach, which permits dividing the monolithic and complex systems into a number of applications consisting of an ecosystem from basic to well-defined components.

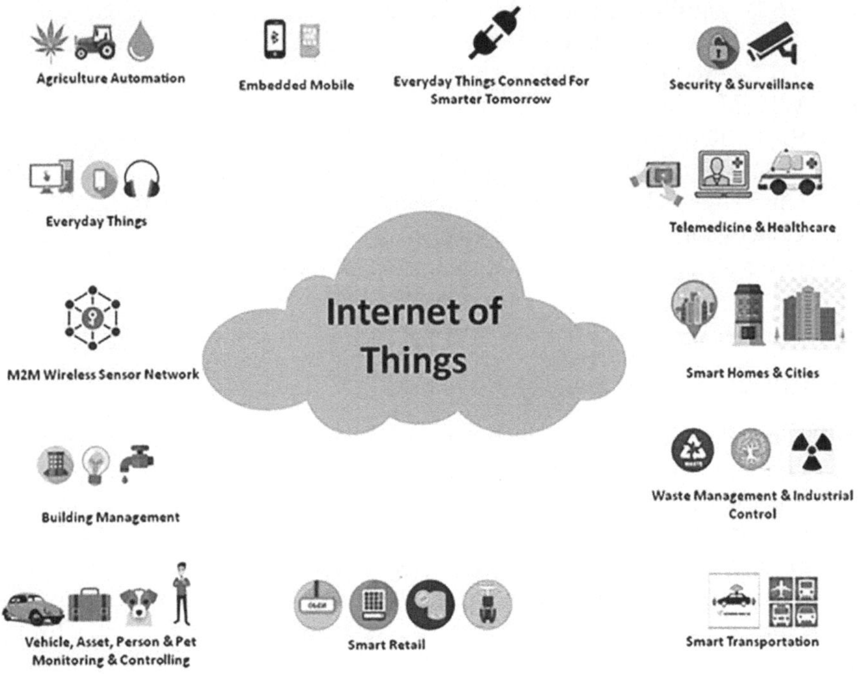

FIGURE 9.1
IOT functioning in different domains.

IoT devices enable smart homes [4,5], such as in the following example: When the user enters their home the user's car communicates with the garage to unlock the door. The thermostat is adjusted to the desired temperature by sensing the proximity of the user. Later, if the user reaches the door, it opens in response to the RFID implant or smart phone. If the home's lighting is attuned to a lower intensity, then the pacemaker data shows that it's been a hectic day. The refrigerator can inform the grocery delivery service that the food essentials, such as soy milk or wine, are low. The component has sensor can communicate to any device that has a sensor. Before waking up completely in the morning, the bed can transmit the information about the sleep quality to the concerned physician.

The IoT can generate new processes and customize existing processes, permitting us to control the secured safe home systems from the office via smartphone or by sending information to a technician that a part of the equipment has failed and the required part needs to be dispatched. For instance, IoT devices are part of any scenario in which every device communicates to every other related device in an environment to automate one's home, industry, etc. and convey huge exploitable data to the business sector, users, and other interested parties. However, IoT refers to an ecosystem of discrete computing devices equipped with sensors with the infrastructure of the Internet. The affordability and availability of miniaturized computing devices has pushed it strongly into the mainstream. Furthermore, the cooperation between such communities is encouraged and foreseen.

9.2 Influence of IoT in the Healthcare Industry

Wearable sensors and devices are smart enough to know the context and monitor our physiology, take particular health-related goals, and alert us to any specific problems are one part ready for ever greater sophistication. Business models are being restructured to adapt to the growing influence of customers in healthcare. The recent developments are due to the extraordinary growth in data which is powered by IoT devices such as wearable and handheld devices [6]. With the Internet of Things, users can acquire the control of their health in an extremely personalized way based on their needs. The main strength of IoT is its data analyzing capacity. Diagnostic information collected using imaging tools amassed through electronic medical annals, monitors screens, and handheld personal devices enriches the decision-making powers of physicians and also empowers patients to take an active role in maintaining their personal health. Apart from supporting the lives of chronically ill people, these innovations also help in transforming the lives of people those who want to stay fit.

An IoT–based, data-rich, personalized study of a person's health [7,8] will become normal by the end of this decade. People would be offered appropriate plans to combat disease and IoT devices that will help them to maintain their own health. With the data generated by many IoT healthcare devices, the significant details can be acquired to improve their health and fight illness. The main challenge of healthcare industry is not just in developing devices and technologies, although the high attention is required in concentrating the IoT healthcare end users. Such devices should be scalable and easy to use. Numerous smart connected devices and solutions aimed to improvise the person's health, productivity, and environment with the intellectual use of data [9,10]. It enables the user to control and monitor air quality and with cloud-based tools. Additionally, the physicians can consult the patients remotely [11].

The IoT is an ever-expanding space of numerous technologies and devices with extensible potential uses. There are three qualities to qualify the sensor equipped "things" to be a part of IoT healthcare [5,12]:

- Initially, the device should be capable of sensing and gathering data about the environment, such as light, temperature (or in the case of blood glucose monitoring, pulse rate/blood oxygen monitoring), electrocardiogram monitoring, etc.
- The device should be autonomous in communicating the gathered data to the central device automatically or with other devices when some conditions are met.
- The device should be in an active condition. For example, if patients' blood sugar or blood pressure levels are at a critical stage, an alert must be automatically triggered to initiate the physician intervention immediately.

Several categories of medical devices [13,14] meet IoT criteria:

- *Consumer-based devices* such as fitness tracking devices such as FitBit, smart watches, etc.
- *Wearable external devices* like insulin pumps;
- *Internally embedded devices* like miniaturized sensors, cardioverter defibrillator devices, and pacemakers
- *Stationary devices* like fetal monitors and home-monitoring devices

For healthcare, the possibilities of applications are tremendously exciting and some healthcare applications are described [15,16]. The irregular heart rate of a patient triggers an alert message to the cardiologist who, in turn, can inform the patient to follow the treatment immediately. Miniaturized implanted devices or skin patches monitor skin temperature and blood sugar and informs the insulin pump to regulate the dosage. Such monitoring helps not only to improve the health status, but also intervenes before the condition becomes critical. Heart failure patients could be equipped with sensors that measure the heart rhythm and tracks them continuously.

Blood pressure, oxygen saturation levels, and medical devices like CTs and MRIs can also be maintained and monitored remotely.

Patient's movements in the hospital can be monitored more efficiently via radio-frequency identification technology and sensor devices on caregivers, using process analytics for identifying and managing the optimal process flow. The person's health is influenced by many behavioral and environmental factors, such as smoking, pollution, etc. The IoT affords data to understand the patient and their own lifestyle, since it has the potential to enhance the effectiveness of health-related activities and strengthening predictive analytics. According to *Gartner's Hype Cycle of Emerging Technologies* report, it would take five to ten years for the full adoption of the IoT by the healthcare sector.

Real-time systems play a significant part in the healthcare IoT, where big data analytic processes and tools are used to assess both static and dynamic data for predictive analysis in healthcare systems. The extensive adoption of emerging technologies, specifically IoT and smart devices, has shown different ways of healthcare delivery for human wellbeing and health. Several applications are now available with more benefits afforded by IoT technologies in healthcare. The steps involved in providing healthcare to patients are depicted in Figure 9.2. It shows the processing steps in healthcare service, includes manufacturing, distribution, patient care, healthcare providers, customer service, and what is expected in the research and development section.

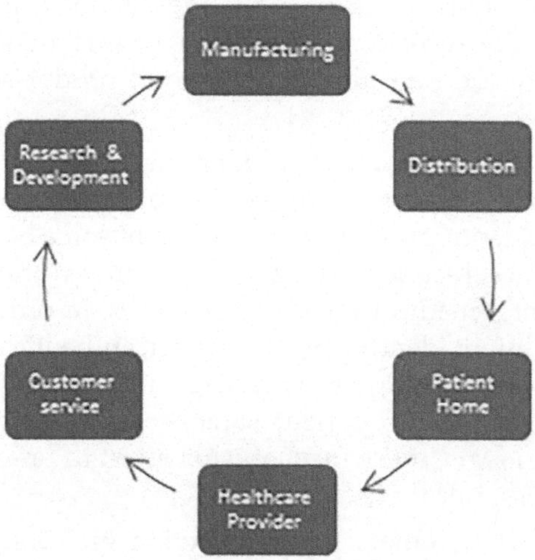

FIGURE 9.2
Healthcare service.

Manufacturing: Equipment should be manufactured using the appropriate technologies to manage them remotely, and machine downtime is also reduced with the help of maintenance alerts.

Distribution: Advanced product monitoring and authentication to prevent unauthorized access should be enabled.

Patient care: The functionality of the medical devices is continuously monitored for reducing the risk, to offer the best in customer service, and insights for better product design. The data would be aggregated and collected from different medical devices for providing advanced insight on health-based outcomes.

Healthcare provider: The significant role of the healthcare provider is to monitor device data consistently for making timely decisions, such as adjusting dosages.

Customer service: Anticipating the maintenance needs of medical devices, and alerting patients to plan a doctor visit for repair or replacement

R&D: Building efficient products informed by a larger data set based on patient outcomes

The IoT applications can be clustered to track objects and people (patients), identify and authenticate users, and automate the data collection and sensing process [17,18].

- *Tracking*: Tracking is the process of identifying the person or object which is in motion. It includes real-time position tracking, monitoring patients continuously and to enrich the process flow in hospitals, and motion tracking via choke points, focusing on designated areas. In the aspect of asset tracking, the idea is often applied in monitoring inventory records, and in terms of materials tracking some shortages like the unavailability of blood products and specimen can be verified.
- *Identification and authentication*: Identification and authentication mechanisms are required in the context of patients, staff, and assets. For patients, identification is vital to prevent some harmful incidents (such as an incorrect dose, drug, time, or procedure) or to maintain the consistent e-medical record for patients, in order to avert mismatching in infant identification. For staff, authentication is predominantly required to assign the privilege and to enrich the employee morale by considering patient safety-related issues. In terms of securing the assets, these mechanisms assist to prevent instrument or product losses or thefts.
- *Data collection*: Automatic data collection and transmission typically targets the reduction of process automation, processing time, auditing procedure, and inventory management. The process also includes the integration of RFID technology with health-related

information and clinical-based technologies which extend the network usage across providers and different locations.

- *Sensing*: Sensor devices gather information about patients by diagnosing the patient condition and the information will be intimated to patient health indicators. The various health-related applications track patient compliance with prescriptions and telemedical solutions, alerting patients if there is a problem. Sensors would be practical in both inpatient and outpatient care. Wireless technologies can be integrated in treatment for supporting constant biosignal monitoring in the case of patient mobility.

Body sensors can be coupled with RasberryPi and Arduino boards to process physiological data using different networks [17,18]. IoT–based healthcare systems are designed to collect biometric information. The information gathered in real time helps to monitor the state of the patient or medical diagnosis can be done with the saved sensitive data, using web services, android applications, and multi-protocol units [19]. The listed elements are essential in IoT–based healthcare systems:

a. Different sensors such as wearable devices, consumer-based, stationary devices, and internally embedded systems
b. Microprocessors, microcontrollers, cloud computing, and the healthcare gateway

IoT–based healthcare [8,20] offers numerous benefits such as accessibility and availability, providing a high-quality and personalized healthcare system. Therefore, the IoT gives promising solutions for healthcare environments; meanwhile patients can manage their own care, enabling secure access for patient-related information, and providing remote assistance via remote monitoring equipment. The life cycle of the patient health treatment process is depicted in Figure 9.3, and is explained below:

1. Continuous patient monitoring starts after scheduling the appointment with the doctor. Initially, patient has to submit complete preliminary health information through the mobile device. The waiting time will be estimated and sent to the patients on their mobile device.
2. The IoT provides a large amount of information for hospitals about their patients, including both inside and outside of the hospital premises. Telemetry in healthcare captures and communicates the data automatically.
3. The diagnostic process has moved far beyond the old technologies and it is obvious that the ideal sources for diagnostic information are now available, such as the molecular contents of saliva, sweat, urine,

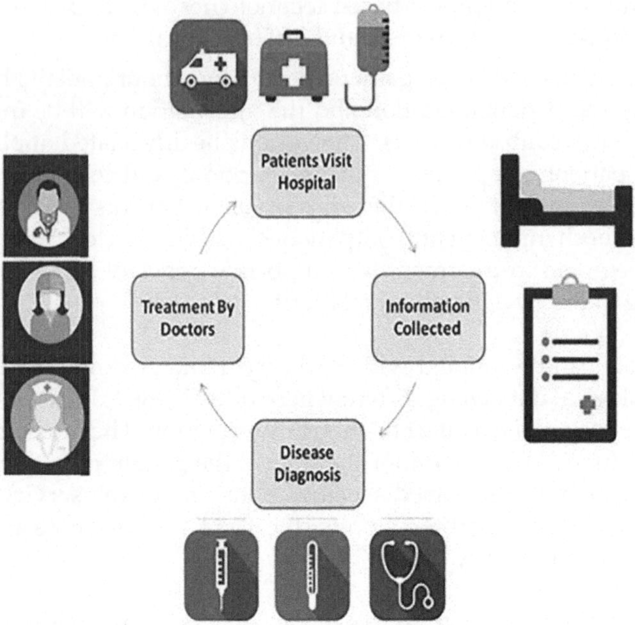

FIGURE 9.3
Lifecycle of the patient health treatment process.

feces, etc. For example, a smart toilet will help to diagnose the level of acidity and glucose in the human body.

4. Doctors can provide an on-time alert message to patients in the moment of life-threatening circumstances. IoT devices collect data and transfer it to doctors, providing treatment by giving notifications to the patient through their mobile devices. Frequent alerts and reports build a strong case regarding the patient's condition. These notifications also assist in providing on-time treatment by making the correct decision and improving the complete patient care delivery system.

The topmost priority of the patient undergoing treatment is to become healthy as soon as possible. Inattentive staff, health record processing delay, and long wait times would significantly delay the patient's access to treatment. With emerging integration of the IoT in daily life, there is an extreme change in hospital care which enables doctors and providers to provide the best possible care.

9.3 Architecture of IoT in Healthcare

The IoT is a network of connected devices with an embedded system to sense, report, and control remote devices. The idea of the IoT [7,8] affords a conceptual framework for connecting edge computing devices (such as wearable sensors) and smart devices for seamless interactions. IoT applications mainly depend upon the information processed in each entity of the middleware layer in the IoT architecture [21]. Some IoT applications are smart health, smart cities, smart homes, smart farming, intelligent transportation, etc.

The basic IoT architecture is a three-layered architecture which consists of perception, network, and application layers [13,22]. IoT architecture has been expanded to have more layered architectures, where the available architecture is not adequate for facing challenges in many IoT applications. Here, five layered IoT architectures are shown Figure 9.4 which include a perception, network, middleware, application, and business layer.

1. *Perception layer*: The perception layer is also called as "device layer", which consists of many sensor devices and physical objects. The sensor devices can be 2D-barcode, infrared, or RFID sensor, based on the object identification method. This layer has sensor devices that identify the object, sense the environment and gather some specific information about it. Other smart objects can also be identified in the same environment by sensing some physical parameters. Based on the type of sensor, the gathered information can be about vibration, location, orientation, motion, temperature, acceleration, chemical changes, humidity in the air, etc. The information is then transmitted to the network layer for secure transmission. Suppose a lady is

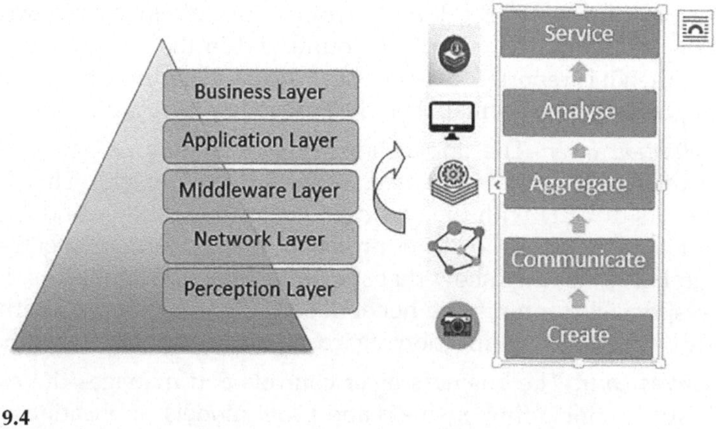

FIGURE 9.4
IoT architectures.

wearing a set of smart earrings in both of her ears. Smart earrings are a wearable sensors, assisting in detecting the physical condition of different parts of the human body and identifying the position of the lady. The perception layer forwards the information gathered by each earring to the network layer [23].

2. *Network layer*: The network layer is also known as the "transmission layer". Its main functionality is to connect various smart objects, network devices, servers. It also safely transmits and processes sensor data which have been collected earlier from sensor devices. The transmission medium can be the wired or wireless, and the technology can be the Bluetooth, infrared, Wi-Fi, ZigBee [24], 3G, UMTS, etc. which is based on sensor devices [25]. Thus, the information has been transmitted from the network layer to the processing layer. The perception layer sends the information gathered by each earring separately. Then network layer combines information from both earrings and forward it to processing layer, i.e., the middleware layer.
3. *Middleware layer*: The middleware layer is the "processing layer." This layer stores, analyzes, and processes the huge amount of data that have been received from the network layer. It is also responsible for the service management and in maintaining the connection with the database. It acts as the middle layer and the devices in the IoT offer a different set of services to the lower layers. The devices can be connected and communicated with the devices which provide the same type of services. It deals with many technologies like cloud computing, big data, and databases for analyzing and processing the large amount of available data. It plays a significant role in ubiquitous computation, information processing, and automatic decision making. The processing layer stores the information gathered from the network layer. Analysis is done on the information gathered by the earrings in order to check body temperature or blood pressure, for example. If any dissimilarity is found within the normal temperature, then it is reported. Suppose that the patient's body temperature is higher than the normal, then the patient has to be notified about it.
4. *Application layer*: The application layer is mainly responsible for delivering application-oriented services to IoT users. This layer interacts directly with the end user, consisting of applications each with its own application layer protocols. If the information gathered from lady's earrings show that she has a fever, the application layer is responsible for notifying her about it. This layer interacts with the lady by sending a notification regarding the fever to her smartphone.
5. *Business layer*: The business layer controls and manages the entire IoT system, including business and profit models, applications, and the user's privacy. This layer deals with the construction of various business models, flowcharts, and graphs depending on the data

obtained from the application layer. The real achievement of the IoT technology lies on efficient and well-constructed business models. After analyzing the result, this layer assists in making decisions concerning upcoming actions and business strategies. As the lady is suffering from fever, this layer will suggest nearby hospitals or clinics, showing the details.

9.4 Wearable Systems

The recent advancement in various domains, such as telecommunication, microelectronics, sensor technology, and data analysis techniques, has paved the way for the development and deployment of wearable systems for patients' health monitoring. Wearables [26–28] are devices that can be fitted on the body in items such as T-shirts, smart rings, watches, pendants, badges, bracelets, glasses, fitness trackers, and other accessories or are surgically implanted in the body. Many users globally gaining the health benefits of wearables. A wearable device which is kept in close contact with the user is capable of monitoring the health and wellness of a person and the gathered data will be sent to a hub station for analysis [29].

Wearable devices [30,31] consist of three components: sensors, computing architecture, and displays. The sensors collect data in an aggregate form about the user; the gathered data will be computed with digital information and displays information which assists in suggesting real-world actions. Wearable gadgets perform basic functionalities like mobile devices, although they are also better in providing users with various biological information, such as heart rate, blood pressure, calories burned, steps walked, and time spent exercising, as shown in Figure 9.5 [32]. The impact of wearable devices is quite powerful, extensive, and gaining attraction, particularly in monitoring user's physical health. However, the implementation of wearable gadgets still been unbeaten like smartphones [31]. All these wearable devices undergone a major weakness as they still unable to interpret a huge amount of information they generate. Once the wearable devices that can arrive with an interpretation of the information generated, these wearables will, in turn, be extremely effective.

Wearable systems are comprised of three major building blocks for patients' remote health monitoring:

1. Sensing and data collection hardware, which are essential to collect the physiological data
2. Communication software and hardware relay data to the remote systems
3. Data analysis techniques that are required to extract relevant information for monitoring the patient's health

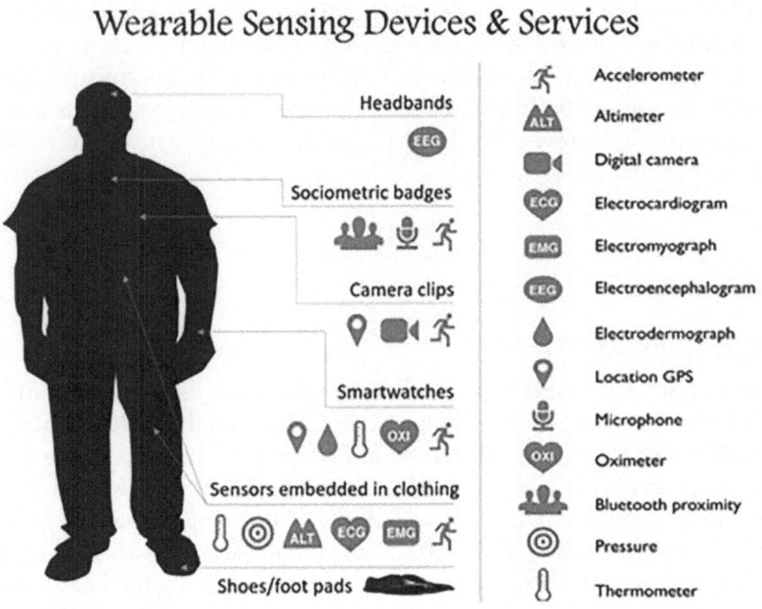

FIGURE 9.5
Wearable devices.

Some wearable devices are described to understand their following applications:

- *Pulse oximetry*: A device that typically measures the oxygen saturation level in a human body and also focuses on the changes in the blood level of skin that is related to cardiac cycle. Normally, a pulse oximeter [33,34] is fixed to an earlobe or a finger, and consists of photodetector and infrared, red light-emitting diodes (LEDs). The photodetector aids to quantify the amount of infrared light that is transmitted to or reflected by the human body, which depends on the amount of light absorbed by the blood in the specific body part. The ratio between the absorption level of the blood and the oxygenated to deoxygenated hemoglobin, helps to measure the oxygen saturation level. The total amount of light absorption is different from the pulsatile volume of blood which changes within the human body part at any time. The periodic signal is known as photoplethysmograph (PPG) that is used to find out the heart rate.
- *Electrocardiography* (ECG): ECG [35] is a waveform that monitors the heart functioning continuously and emphasizes the information with the electric potentials with respect to time. Hence, ECG waveform affords a noninvasive method of investigating heart functionality. Standard ECG measurement uses 12 leads

or "views" of the electric potential activity of the heart. Moreover, ECG measurements using wireless sensor devices are relatively suitable for ambulatory applications and will typically use a subset of these leads.

- *Blood pressure*: A blood pressure (BP) reading helps to measure the force exerted due to the circulation of the blood on the walls of blood vessels [35,36]. A BP sensor can be fixed around the wrist using an inflatable cuff and the systolic and diastolic readings are measured using the oscillometric method. However, this method can be utilized to capture ambulatory BP readings, the BP variations cannot be monitored continuously and the cuff-based measurement may not always be comfortable for users.
- *Electromyography* (EMG): An EMG is the study of muscle function through the continuous monitoring of the electrical signals exerted by the muscle. Hence, the EMG signal is the spatiotemporal summation of resultant electrical signals. Therefore, the EMG signal offers an effective means of monitoring muscle activity.
- *Activity/motion detection*: The nature of motion or the level of activity of an individual can be sensed by a system that merges an accelerometer with a gyroscope. An accelerometer is a sensor which measures the acceleration with respect to gravity and can be utilized to govern the orientation of a body part in the lack of movement. A gyroscope is a sensor which measures the angular velocity and can be utilized to govern the orientation of a moving body part with respect to time.
- *Electroencephalography* (EEG): EEG is the illustration of electrical activity of the human brain. Wireless intelligent sensors (WISE) for EEG signal acquisition applications are introduced and it is a microcontroller-based system capable of analog–signal conditioning, data acquisition, wireless communication, and low-level, real-time signal processing.

9.5 Implantable Systems

Implants [37,38] are devices that can be implanted beneath the skin of the human body and assist in replacing or enhancing the part of or entire biological structure. Recently, implants are normally used in different parts of the body for a variety of applications like cardiovascular stents, artificial pacemakers, neural prosthetics, orthopedics, defibrillators, drug delivery systems, cochlear implants, and dental fillings and crowns, as depicted in Figure 9.6.

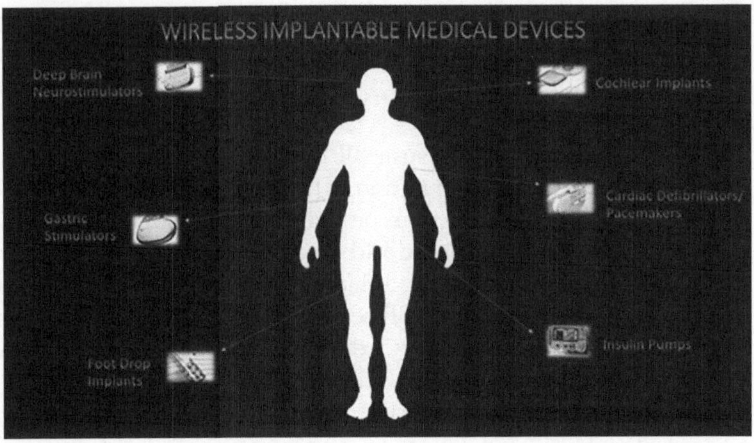

FIGURE 9.6
Implantable devices.

The outer surface of implant might be made of any biomedical material such as silicone, titanium, or apatite. The material should be chosen based on its functionality, since it is in close contact with the human body.

- The materials used for implant devices can be categorized as metals, ceramics, or polymers. The classification of medical devices depends on the level of control required to guarantee the safety and effectiveness of the device. Metals are based on the metallic bond, where the metallic materials consists of pure metals and alloys; polymers are based on covalent bonds, where the polymers consists of thermoplastics, thermosets, textiles, and elastomers; and ceramics are based on ionic bonds which include carbons, glasses, and glass-ceramics.
- Implantable biosensors characterize a very desirable proposition for diabetes management which presently relies on data obtained by means of test strip blood from finger pricking, a practice which is not only painful, but also is incapable of reflecting the whole direction, pattern, and trends interrelated with daily habits.
- The vision of the blind person can be improved by implanting sensor devices in human eyes and it is viewed as the significant application in healthcare domain. Patients who are visually impaired or have no vision at all can see at a reasonable level by implanting retina prosthesis chips within their eyes.

Some of the implantable medical devices are quoted in Table 9.1.

According to experts and the expectations and predictions of E-marketers, the usage of wearable gadgets in 2018 would reach 81.7 million from

TABLE 9.1
Classification of Implantable Medical Devices

Implantable Types	Cardiovascular	Spinal	Orthopaedic	Cardiac Stents	Otolaryngeal
1	Pacing devices	Thoracolumbar implants	Reconstructive joint replacements	Coronary stents	Cochlear implants
2	Cardiac resynchronization devices	Intervertebral spacers	Hip replacement	Stent related implants	Airway and oesophageal stents
3	Implantable cardioverter	Motion preservation devices	Shoulder replacement		Ophthalmic implants
4	Implantable cardiac pacemakers	Implantable spinal stimulators	Elbow replacement		Gastroenterological implants
5	Pacing accessories	Cervical implants	Ankle replacements		

39.5 million in 2015. As the message highlighted at World Economic Forum, implantable technology and some other new technologies can be anticipated. The expected new trends by 2023 may be human vision interface, pocket super computer and the implantable technology is one of them. For instance, it is also quite possible that wearable devices would be surpassed by implantables by 2023, meaning smart devices that can be implanted beneath the skin for tracking health-related information. The implantable gadgets will definitely revolutionize the current smart technologies if the following features are well-supported:

- ICs should be inexpensive for manufacturing huge quantities of electronic devices.
- Microscopically small features make ICs a very effective method for reducing the size, weight, and power of electronic devices.

Some implantable devices are discussed as follows:

- *Glucose Monitoring*: Continuous glucose monitoring can be done by fixing the implantable sensor device wrapped with the multilayered membrane in the abdominal tissue. Body glucose level can be determined for every 30 s and the glucose data transmission occurred for every 5 minutes. If the sensors are embedded with the implantable drug delivery system, the blood glucose level can be controlled by delivering the variable amount of insulin, hence the closed feedback loop is formed.
- *Implantable Neural Stimulators*: Implantable neural stimulators direct electrical impulses into the spinal cord or the human brain to afford treatment for intractable epilepsy, Parkinson's disease, and chronic pain.

9.6 Wireless Body Sensor Networks (WBSNs)

WBSNs [31] have the potential to provide a huge number of benefits to medical staff, patients, and society through constant monitoring in an ambulatory mode, supervised rehabilitation, early detection of abnormal disorders, and discovering knowledge about the patient from the gathered information. The devices are located throughout the network, as the deployment of device depends on a certain application. The spot of sensor nodes is not fixed as the human body changes its position always. Therefore, WBSNs cannot be viewed as a fixed network. The communication design [39,40] of WBSNs comprise three levels that are shown in Figure 9.7.

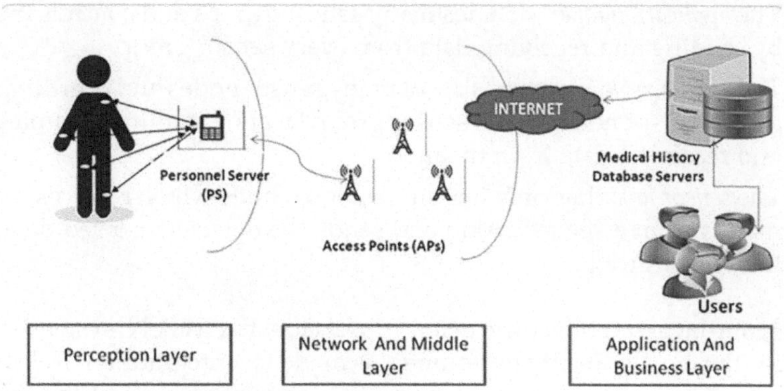

FIGURE 9.7
WBSN communication architecture.

Tier 1: *Intra-WBSN communication* is limited around the proximity of human body. The communication signals within the specified area exploits a personnel server (PS) which acts as a gateway and also it transmits the information to the access point (AP) in the next level (Tier 2).

Tier 2: *Inter-WBSN communication* bridges the gap between the personnel server and access point which plays a significant role in the network and that may be located to allow the emergency cases. Connecting WBSNs with other networks enables the transmission of information easily via different mediums.

The architecture of inter–WBSN communication [40] is classified into two subgroups, infrastructure- and ad hoc–based:

a. Infrastructure-based architecture is widely used in WBSN applications. It permits the active utilization, even in a constrained area. This model also helps to achieve a better security level and central management, since the access point acts like a database server.

b. Ad hoc–based architecture has many access points are designated to relay information by creating a lattice that allows flexible and quick disposition. Additional radio coverage could be achieved via expansion and nonlinear distribution. The configuration of ad hoc–based architecture is larger when compared the infrastructure-based coverage. Also, ad hoc–based architecture extends the coverage area from 2 m to 100 m.

There are four levels in WBSN architecture, namely:

1. The *system layer* consists of the data path and the control path for WBSN system. The control path consists of power control and function control signals. The main responsibility of this layer is to send commands to the application later and to receive data from it.

2. The *application layer* includes many sensor groups and controls them by sending and receiving data from every sensor group.
3. The *sensor group layer* consists of many sensor nodes and coordinates all the sensor nodes in the sensor group layer by sending commands and receiving data from them.
4. The *sensor layer* has only one single sensor node which receives commands from a sensor group and sends the detected, sensed data to the sensor group.

The four-layer architecture helps to develop the WBSN system [41,42]. Initially, the development environment is made for integrating wireless sensor nodes. The sensor group is viewed as a geometric node of the Voronoi diagram, which permits sensor node to check the connection between sensor groups and sensor nodes. Then, an application layer is divided into number of cells, which facilitates communication when the sensor nodes are in mobility. In WBSNs, each sensor node should be assigned with a unique identification code and the sensor groups are designed as the geometric nodes of a Voronoi diagram, as mentioned earlier. It also ensures that the biomedical data can be delivered properly at the end. Moreover, the WBSN system can be incorporated into a geographic or metropolitan network for online healthcare-monitoring systems.

9.6.1 Significance of the Medium Access Control Layer in WBSNs

The medium access control (MAC) layer is the sublayer of the data link layer and it is the second layer in the open system interconnection (OSI) model. The main responsibilities of the MAC layer include data framing, transmission scheduling, channel access control, energy management, and error handling. Accordingly, the MAC layer is the appropriate level for managing energy efficiency–based issues. Different MAC protocols have been introduced in order to reduce energy consumption and some reasons for energy consumption in wireless network [43–47] are listed below:

- *Packet collision*: Packets will be collided if more than one packet is transmitted simultaneously. The packets which are collided need additional energy consumption for retransmitting those packets again.
- *Idle listening*: If and when a node is waiting for an idle channel to collect the data
- *Overhearing*: When a node is listening for the channel to accept the packets that are meant for some other nodes
- *Packet overhead*: When the number of control packets is added to carry some additional information with the data communication process, which also raises energy consumption.

9.6.2 Requirements of MAC Protocol in WIBSNs

Some of the basic requirements of a MAC protocol are described below. They are the size, amount of data flow, energy efficiency, latency, fairness, adaptability, and network density [14,40].

- *Energy efficiency*: With a huge number of powered battery nodes, the process of upgrading or recharging batteries of those nodes is very difficult.
- *Adaptability*: As many nodes have entered or left the network frequently, the protocols used must be able to manage them dynamically. The nodes must be programmable in such a way that if new tasks arrive, the operations should be handled without any interruption.
- *Latency*: The monitoring nodes in wireless body area network (WBAN) will be attentive for a long time, although the network is highly inactive until some anomalies are detected. Low latency can be achieved by reducing the consumption due to the limited data flow in the network.
- The *amount of data flow* refers the amount of data transmitted from a sender to a receiver in a particular period of time. The factors which are affecting the data rate include the usage of channel, latency, effectiveness of collision avoidance.
- *Fairness*: The number of nodes in the network should share or utilize the channel in a distributed manner. Although, in WBAN, every node requires an equal attention on sending or receiving the data. A node may or may not have the data to transmit, rather the success rate is measured based on the overall performance of the network.
- *Network density*: The densities of nodes in the network will be different. They may vary according to space and time based on the tendency of active or inactive mode. The density across the network is also not homogeneous and the network should be adaptable to these modifications.

9.6.3 Energy Efficient Routing Protocols for WBSNs

In wireless communication, sensor nodes are either placed inside the human body or around the human body, causing different types of effects on the body. The most significant factor for this purpose is specific absorption rate (SAR) and the heating effects of the implanted sensor nodes on the human body. As sensor nodes have limited energy sources, they require frequent recharging, which is very important factor for the network's lifetime. Energy preservation is a prime consideration in WBSNs [48,49]. Multi-hop communication is used to increase energy efficiency and the network lifetime of a routing protocol. Basically, routing protocols are broadly classified into five categories based on design and constraints: temperature-based, cluster-based, cross-layer, cost-effective, and QoS–based [50].

The main phases of communication in routing protocols:

- The *initialization phase*, in which short messages are broadcasted by all nodes which contain neighbor's information and distance of sink nodes in the form of hop-counts. In this way, all nodes are updated with their neighbors, sink position and available routes to the sink node.
- The *routing phase*, which is responsible for selecting the routes with fewer hops to sink to minimize the distance and consume less energy. Emergency facilities to select the fastest route should be incorporated in this phase.
- The *scheduling phase*, where sink nodes allocate time-slots to nodes. Nodes can communicate to sink node in the assigned time slot for normal data delivery.
- The *data transmission phase*; once the time slots are allocated to root nodes which can send their data to sink node in the assigned time slot. After receiving data from the sink node, it takes some time to aggregate the received data.

The crucial performance parameters for the existing routing protocols in WBSNs are the network stability period, residual energy, the network lifetime, throughput, delay, and path loss [51,52]. The temperature-based routing protocols applied for in-body body sensor networks (BSNs) consider temperature as a factor for choosing routes that would either avert hot regions or find a detour after reaching a hot region. The thermal-aware routing algorithm (TARA) [53] protocol uses sensor locations and cluster leadership history to minimize the hazardous effects of temperature rise on the human tissue. Bag et al. [54] have proposed the least temperature routing (LTR) protocol which is a thermally aware routing protocol for BANs. LTR defines hot spots as areas which have high temperature due to data communication focus. Unlike TARA, LTR chooses its routes from neighbor nodes with the lowest temperature. Hence, it sets its path to the coolest neighbor without involving routing loops. LTR is considered a greedy approach to routing that is not globally optimal. Nabi et al. [55] proposed a multi-hop protocol to monitor human body health conditions. This protocol is robust when subjected to frequent network topology changes due to the variation in wireless link quality and postural changes [56,57].

Javaid et al. [58,59] presented a thermally aware routing protocol in which each node selects a minimum hop route to the sink. Moreover, this protocol uses both single hop and multi-hop communication modes by choosing the former one for emergency data cases and the latter one for normal cases.

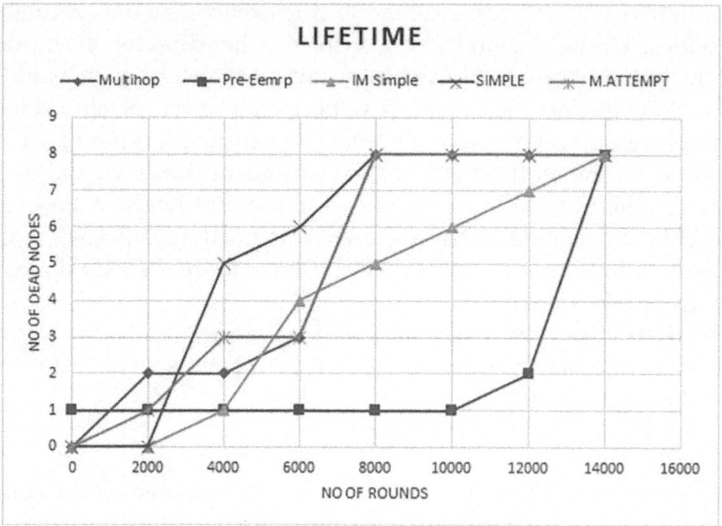

FIGURE 9.8
Comparison of different routing protocols.

In this way, the emergency data are sent to sink with the minimum possible delay. The cost-effective routing protocols compute the probability of a link to be connected, depending on the knowledge from certain characteristics in the network [51]. However, no cost-effective routing protocols consider temperature increase in the nodes and the path loss happening in sensor nodes around the human body. The cross-layer and cluster-based routing protocols are primarily reactive and need to acquire knowledge of the connectivity of all nodes in the network and their other features which leads to significant overhead. QoS–based routing protocols aim to accomplish the required QoS metrics. Javaid et al. [58,59] proposed a new multi-hop routing protocol improved stable increased throughput multi-hop link efficient routing protocol for wireless body area networks (which is extended form of the SIMPLE [59] protocol), which supports mobility and selects a forwarder node on the basis of cost function. iM-SIMPLE [59] attained good performance when compared to other existing schemes by considering the throughput and maximizing stability period of the network. The comparison of different routing protocols in WBSN [55] is given in Figure 9.8.

9.7 WIBSN in Healthcare

WIBSN [60–62] uses embedded sensors for monitoring the patient's health continuously and to collect health-related information. It is comprised of a wireless network which contains many tiny devices such as actuators and sensor nodes. The sensor nodes can be fixed either directly on or under the skin of the human body to figure out certain parameters, such as EEG, temperature, blood glucose, heart rate, ECG, body movement, plasmon biosensor, blood pressure, and respiration rate levels. The different types of sensors are devised to meet the requirements of the various purposes of end users [40]. For instance, the ECG sensor is intended to monitor heart related activities and the EEG is designed to track brain-related activities [63,64]. According to the taxonomical structure of IEEE 802.15.6 standard, WBAN nodes can be fixed within the human body.

This section defines how to use the mechanism and architecture of WIBSNs [27,28,30] for enabling the daily monitoring of patients' health status and is illustrated in Figure 9.9.

The categories of WIBSN nodes are:

- *Implant Node*: This is the node which can be planted either beneath the skin or inside the tissue of a human body.
- *Body Surface Node*: This type of node can be either placed 2 cm away from the human body or on the surface of it.
- *External Node*: This is the node that is not in a direct contact with the human body and slightly a few cm to 5 m away from the human body.

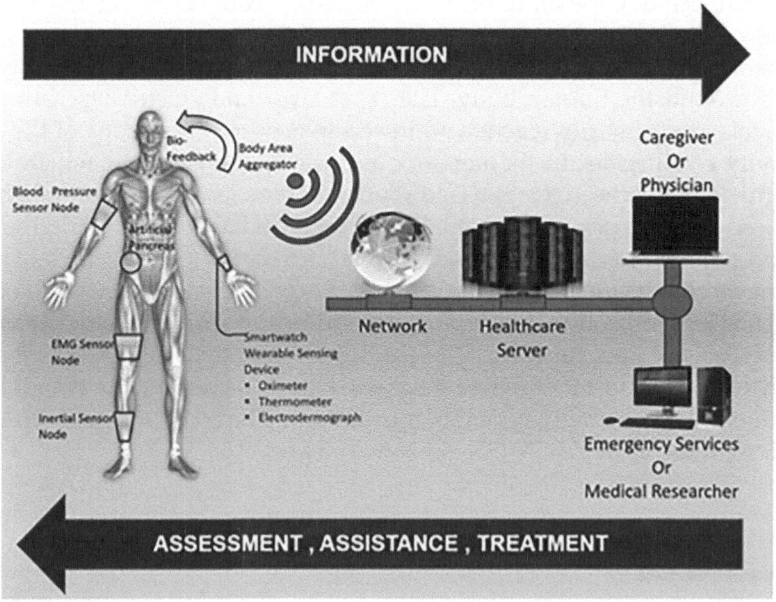

FIGURE 9.9
WIBSN architecture in healthcare.

The different roles of three nodes in WIBSN are:

- *Coordinator*: This node acts as an access coordinator or a trust center and also viewed as a gateway to the outside world.
- *PDA*: The coordinator which enables all remaining nodes to communicate with each other
- *End nodes*: These nodes are not allowed to transmit messages from one node to another node.

The intermediate nodes are known as relays. They consist of a parent node, a child node, and relay messages. If a node is at the bottom, the data is relayed by other nodes before reaching the PDA. These nodes can also sense data from some other nodes. Actuators will react based on the information received from sensors depending on prescribed instructions. The actuator mechanism is ready with integral reservoir and administers the proper doses of hypoglycemic agents to support the glucose level measurement in diabetic patients [65].

9.7.1 Requirements Needed for Medical Sensors in WIBSNs

Wireless medical sensor devices should satisfy the following requirements: wearability, reliability, interoperability, and security [66,67].

- *Wearability*: To attain noninvasive and unobtrusive continuous health monitoring, wireless medical sensor devices must be tiny and lightweight. The weight and size of sensors are generally determined by the weight and size of batteries. Henceforth, the advancement of the miniaturization of batteries and integrated circuits improve the wearability ratio and comfort of patients.
- *Reliable Communication*: Both computation and communication should be considered for optimal system designing. Different medical sensor devices are required for communication that can be chosen based on sampling rates ranges from 1 to 1000 Hz. The proper processing of sensor signals improves the reliability during communication. In order to face the demands of the communication channel, the communication devices should be chosen carefully based on saving the energy expenditure and increasing the battery life consequently.
- *Interoperability*: Medical sensor devices should be made available and easy to construct a robust network based on the user's health status. Interfaces govern the interaction among sensor devices and will assist vendor competition and ultimately lead to have more accessible systems.

- *Security*: Security is the significant issue in the complete system of WBSNs. Security issues should be considered in all levels of WBSN-based healthcare systems. Basically, every medical sensor device should ensure privacy and data integrity. With the key establishment, data integrity and authentication are challenging with limited resources of sensor devices, relatively short communication ranges, and a fewer number of nodes that makes these tasks attainable.

9.7.2 Parameters for Designing IoT–Based WIBSNs

User-driven design methodology (DM) is proposed to solve the problems which occur in real-life scenarios for supporting persons with limited abilities, especially for senior people living alone. The design system consists of the inertial measurement unit (IMU) with an integrated accelerometer, magnetometer, altimeter, and gyroscope, along with Wi-Fi, heart rate modules, predefined threshold, decision tree, and pedestrian dead reckoning (PDR) algorithms [68–70] that are working properly in the real time environment. In IoT healthcare applications, the most predominant issues are the localization of the equipment or patient [71–73]. Many existing approaches depend on received signal strength (RSS). The appropriate algorithms and technologies should be selected based on the necessity of healthcare applications and can be carried out from the set of probable solutions for basic defined constraints and functionalities. Some of the technologies and algorithms are listed in Table 9.2.

The PDR algorithm [70], which applies three-axial accelerometer and magnetometer data, is selected for an indoor localization with four room-level accuracy in an apartment. This recommendation, in addition to meet the room level constraints, maintains comfortable use of the system. Moreover, the simplicity of the PDR algorithm enables the comfortable communication between the PC and the designed device. For an indoor localization in a multi-story building at floor level resolution, the BarFi algorithm [74] based on fingerprints of atmospheric pressure measurements and Wi-Fi signal, is

TABLE 9.2
Parameters for Designing WIBSN

Certain Parameters for WIBSN Design	Technologies and Algorithms Involved in Designing
Localization	Bluetooth, PDR, finger prints, GPS, GSM, RFID.
Activity monitoring and recognition	RFID, decision trees, genetic algorithms, neural networks, accelerometer.
Vital signs monitoring	Infrared, acoustic, electrocardiogram.
Behavior prediction and classification	Genetic algorithms, decision trees, k-nearest neighborhood.

chosen. The reason is that of simplicity, which ensures comfort during communication. In addition to that, due to its versatile nature, it satisfies the convenience of usage constraint.

The PDR and GPS-based hybrid methodologies are suitable to provide accuracy for at least 10 m in outdoor locations. The common design constraints of the system (such as low energy consumption, easy use, low-cost installation, and small size) are supported by using Arduino technology along with some compatible devices. The durability of the system can be ensured by the following energy-saving adaptive algorithms; for instance, the localization sampling is adjusted with respect to the actual subject position. The combined set of technologies is adequate for the identification of the subject's dissimilar postures and activities.

Behavior prediction and classification: If there is a change in the behavior, fall detection, or localization of the monitored person, there is a need to discriminate the person's posture, like walking, lying, sitting, or standing. It is even required that the system should distinguish the recent behavior of the monitored person is normal, suspicious, or dangerous. In case, the unusual activities are identified and the concerned person should be notified for taking care immediately. Moreover, if behavior is suspected as suspicious or dangerous, the information should be afforded with some significant signs for that recognized subsidiary part.

9.8 Challenges, Open Research Problems, and the Future of WIBSNs

Some observed challenges [2,3,7] in design-pervasive healthcare systems are presented and open research problems of the wearable and implantable sensor systems are discussed. There are numerous challenges of WIBSNs in various perspectives:

a. *Unobtrusiveness*: The great challenge lies in the design and development of wearable implantable sensor devices without violating unobtrusiveness. If the sensors are embedded with human bodies and the patients are carrying always as discussed in fall detection system, unobtrusiveness is considered as the serious problem while comparing other challenges. The integration of different sensors into a single solution creates both the sensor units of LiveNet and paths. The body-worn sensor devices are heavy and intrusive, whereas the watch-shaped activity recorder and the bandage type electrocardiogram sensors are easier to wear.

b. *Sensitivity of sensors*: The sensitivity of the sensor devices is especially significant when users wear these sensors in hazardous environments, such as in fire situations. Sweat generated in the human body can disturb the transducers of the sensor devices negatively which causes a reduction in the sensitivity of the body-worn sensors or demands the recalibration of sensors. An automatic self-calibration algorithm is also proposed for triaxial accelerometers. However, the sensitivity enhancement and self-calibration algorithms are still required for sensor devices unlike than accelerometers. Highly sensitive and low maintenance monitoring sensors would gain the significance as the healthcare systems evolve.

c. *Energy of batteries*: The conservation of energy is considered as one of the crucial challenges in WBSNs and hardly requires distinct investigations and solutions. Frequent recharge of batteries in multiple sensors is another issue yet to be resolved. Robust energy conserving techniques related to the term "green technology" are still required, although lot of effort is made to design low-power sensors. Hence, body heat and motion-based energy techniques should be designed for healthcare systems.

d. *Effective methods for data collection*: The data collection rate in widespread healthcare systems and pervasive WSNs is high. The exploration in finding innovative methodologies for data processing techniques are considered to be important issue. In a few cases 3-lead electrocardiograms may not be adequate to identify the heart diseases or a single 3-axes accelerometer cannot categorize all activities of people. In such cases, additional sensors are required, and the need for the data collection process will also increase. The data collection and analysis of real time physiological data is indispensible. In addition, achieving synchronization of different sensors, time-stamping, and the ordering of events are open problems in this area of research. Finally, modular architectures are required due to the integration of different types of sensors, including wireless body sensors, implantable sensors, RFID tags, etc.

f. *Compatibility problem*: The integration of multiple sensor devices that function at different frequencies raises compatibility problems. Communication between those devices fetches multiple bands and utilizes different protocols. This will lead to more interference among different devices, especially in unlicensed scientific, industrial, and medical band radio. The healthcare systems and WBSNs must be developed in such a way that assures compatibility between the different devices.

Significant improvements in the design and characteristics are required for the benefit of WBSNs in the future. The factors which influence the design of WBSN IoT-based systems are described below:

- Low size and low expensive sensor node electronics designed with wireless capability are vital. Sensor nodes must be capable of transmitting data at a distance of several meters. The sensor nodes must be miniaturized to make the wearables or implantables comfortable.
- Physiological data must be categorized as crucial as well as non-crucial data for each patient. High-gain miniatured antennas should be developed for sensor nodes to minimize interference and to increase the transmission reliability, thus reducing energy consumption.
- The transmission mechanism should be embedded into the WBSN, which enables the patients to move freely in medical applications. These can be used to monitor patients continuously inside a hospital or the patient can be tracked out when they are going about their daily activities outside the premises.
- Personal health monitoring systems are only used for collecting the data. The data analysis and processing are accomplished offline, which is impractical for the early detection of health disorders and for continuous monitoring.
- Individual sensors should be able to operate as standalone systems in WBSNs, and typically integration with third-party devices must not be encouraged.

9.9 Conclusion

WIBSNs provide a technological infrastructure that uses wearable health sensors to facilitate the monitoring of human factors such as wellness, health, behavior, and other chronic data beneficial in enriching individual's quality of life. WIBSN targets connecting the body-worn sensor devices to the medical domain such that physicians can supervise the health condition of their patients remotely. The recent explosion of personal computing devices into the consumer market that combine social networking applications can be boosted with the introduction of WBSN without the active involvement of the individual at the center of these networks. This kind of passive involvement in data transfer can ease the cognitive burden on the individual and result in more unobtrusive computing applications. It also has the potential for breakthroughs in the study of medicine. The motive of Chapter 9 is to examine the various attributes associated with wearable and implantable body sensor networks and analyzing its potential benefits with respect to healthcare. From this chapter 9, the reader will get an overall perspective of the entire ecosystem of IoT–based healthcare and functioning of the components associated with it.

References

1. S. M. Riazul Islam, Daehan Kwak, Md. Humaun Kabir, Mahmud Hossain, Kyung-Sup Kwak. "The Internet Of Things For Health Care: A Comprehensive Survey." *IEEE Access*, Jun 2015.
2. J. Holler, V. Tsiatsis, C. Mulligan, S. Karnouskos, S. Avesand, D. Boyle. *From Machine-to-Machine to the Internet of Things: Introduction to a New Age of Intelligence*. Amsterdam, The Netherlands: Elsevier, 2014.
3. Z. Pang. "Technologies and architectures of the Internet-of-Things (IoT) for health and well-being." M.S. thesis, Dept. Electron Computing System, KTH-Roy. Inst. Technol., Stockholm, Sweden, Jan 2013.
4. L. Tan, N. Wang. "Future Internet: The Internet of Things." *Proc. 3rd Int. Conf. Adv. Comput. Theory Eng. (ICACTE)* Vol. 5, V5-376–V5-380, Aug 2010.
5. J. Ko, C. Lu, M. B. Srivastava, J. A. Stankovic, A. Terzis, M. Welsh. "Wireless sensor networks for healthcare." *Proc. IEEE*, Vol. 98, No. 11, 1947–1960, Nov 2010.
6. Dey, N., Ashour, A. S., Shi, F., Fong, S. J., Sherratt, R. S. "Developing residential wireless sensor networks for ECG healthcare monitoring." *IEEE Transactions on Consumer Electronics* 63(4), 442–449, 2017.
7. W. Wang, J. Li, L. Wang, W. Zhao. "The Internet of Things for resident health information service platform research." *Proc. IET Int. Conf. Commun. Technol. Appl. (ICCTA)* 631–635, Oct 2011.
8. Z. Pang, Q. Chen, J. Tian, L. Zheng, E. Dubrova. "Ecosystem analysis in the design of open platform-based in-home healthcare terminals towards the Internet-of-Things." *Proc. Int. Conf. Adv. Commun. Technol. (ICACT)* 529–534, Jan 2013.
9. Dey, N., Hassanien, A. E., Bhatt, C., Ashour, A., Satapathy, S. C. "Internet of Things and Big Data Analytics Toward Next-Generation Intelligence." *Springer* 2018.
10. Bhatt, C., Dey, N., Ashour, A. S. "Internet of Things and big data technologies for next generation healthcare." 2017.
11. Y. J. Fan, Y. H. Yin, L. D. Xu, Y. Zeng, F. Wu. "IoT-based smart rehabilitation system." *IEEE Trans. Ind. Informat.* Vol. 10, No. 2, 1568–1577, May 2014.
12. H. Alemdar, C. Ersoy. "Wireless sensor networks for healthcare: A survey." *Computing. Network* Vol. 54, No. 15, 2688–2710, Oct 2010.
13. Ahmad, M. "Designing for the Internet of Things: A paradigm shift in reliability." *Proceedings of the 2015 IEEE 65th Electronic Components and Technology Conference* (ECTC), San Diego, CA, 26–29, pp. 1758–1766, May 2015.
14. Dey, Nilanjan, Amira S. Ashour, Fuqian Shi and R. Simon Sherratt. "Wireless Capsule Gastrointestinal Endoscopy: Direction-of-Arrival Estimation Based Localization Survey." *IEEE Reviews in Biomedical Engineering* 10 (2017): 2–11.

15 P. Rocca, A. F. Morabito. "Optimal synthesis of reconfigurable planar arrays with simplified architectures for monopulse radar applications." *IEEE Trans. Antennas Propag.* Vol. 63, No. 3, 1048–1058, 2015.
16. M. M. Sandhu, N. Javaid, M. Jamil, Z. A. Khan, M. Imran, M. Ilahi, M. A. Khan. "Modeling mobility and psychological stress based human postural changes in wireless body area networks." *Computers in Human Behavior,* 2014.
17. K. Vasanth, J. Sbert. "Creating solutions for health through technology innovation." Texas Instruments. Available: http://www.ti.com/lit/wp/sszy006/sszy006.pdf, accessed Dec. 7, 2014.
18. L. Yang, Y. Ge, W. Li, W. Rao, W. Shen. "A home mobile healthcare system for wheelchair users." *Proc. IEEE Int. Conf. Comput. Supported Cooperat. Work Design* (CSCWD), 609–614, May 2014.
19. Wannenburg, J., Malekian, R. "Body sensor network for mobile health monitoring, a diagnosis and anticipating system." *IEEE Sens. J.* Vol. 15, 6839–6852, 2015.
20. Saini, A., Yammiyavar, P. "Weak eyesight therapy: A case study in designing an application for m-health systems." *Proceedings of the 2013 International Conference on Human Computer Interactions* (ICHCI), Chennai, India, 23–24, pp. 1–8, Aug 2013.
21. X. Wang, J. T. Wang, X. Zhang, J. Song, "A multiple communication standards compatible IoT system for medical usage." *Proc. IEEE Faible Tension Faible Consommation* (FTFC), pp. 1–4, Jun 2013.
22. Coetzee, L., Eksteen, J. "The Internet of Things—Promise for the future? An introduction." *Proceedings of the 2011 IST-Africa Conference Proceedings,* Gaborone, Botswana, 11–13, pp. 1–9, May 2011.
23. Mainetti, L., Patrono, L., Vilei, A. "Evolution of wireless sensor networks towards the Internet of Things: A survey." *Proceedings of the SoftCOM 2011, 19th International Conference on Software, Telecommunications and Computer Networks,* Split, Croatia, 15–17, pp. 1–6, Sep 2011.
24. Liu, X. Q., Cai, W. M. "The alarm system of elder tumble at the geracomium based on ZigBee." *Proceedings of the 2011 International Conference on Electronics and Optoelectronics* (ICEOE), Dalian, China, 29–31, pp. 38–40, Jul 2011.
25. Khalil, N., Abid, M.R., Benhaddou, D., Gerndt, M. "Wireless sensors networks for Internet of Things." *Proceedings of the 2014 IEEE Ninth International Conference on Intelligent Sensors, Sensor Networks and Information Processing* (ISSNIP), Singapore, 21–24, pp. 1–6, Apr 2014.
26. M. Soini et al. "Wireless Body Area Network for Hip Rehabilitation System." *Ubiquitous Computing and Communication Journal* Vol. 3, 1–7, 2008.
27. E. Jovanov et al. "A wireless body area network of intelligent motion sensors for computer assisted physical rehabilitation." *Journal of Neuro Engineering and Rehabilitation* Vol. 2, Mar 1, 2005.
28. Ullah, S., Higgins, H., Braem, B., Latre, B., Blondia, C., Moerman, I….Kwak, K. S. "A comprehensive survey of wireless body area networks." *Journal of Medical Systems* Vol. 36, No. 3, 1065–1094, 2012.
29. Rodrigues, J. J., Pereira, O. R., Neves, P. A. "Biofeedback data visualization for body sensor networks." *Journal of Network and Computer Applications* 34(1), pp. 151–158, 2011.
30. Tobon, D. P., Falk, T. H., Maier, M. "Context awareness in WBANs: A survey on medical and non-medical applications." *Wireless Communications,* IEEE Vol. 20, No. 4, 30–37, 2013.

31. Boulemtafes, A., Badache, N. "Design of Wearable Health Monitoring Systems: An Overview of Techniques and Technologies." *mHealth Ecosystems and Social Networks in Healthcare* 79–94, 2016.
32. Gjoreski, H., Lustrek, M., Gams, M. "Accelerometer placement for posture recognition and fall detection." *Proceedings of the 2011 7th International Conference on Intelligent Environments* (IE), Dallas, TX, USA, 19–22, pp. 47–54, Oct 2011.
33. Jovanov, E., Milenkovic, A. "Body area networks for ubiquitous healthcare applications: Opportunities and challenges." *Journal of Medical Systems* Vol. 35, No. 5, 1245–1254, 2011.
34. Li, H. B., Takizawa, K., Kohno, R. "Trends and standardization of body area network (BAN) for medical healthcare." EuWiT. IEEE European Conference, 1–4, Oct 2008.
35. Benoit Latre, Bart Braem, Ingrid Moerman, Chris Blondia, Piet Demeester. "A Survey on Wireless Body Area Networks." *Wireless Network* Vol. 17, 1–18, 2010.
36. Mehdi Effatparvar, Mehdi Dehghan, Amir Masoud Rahmani. "A comprehensive survey of energy aware routing protocols in wireless body area sensor networks." *J Med Sys.*, 2016.
37. Pervez Khan, M., Hussain, A., Kwak, K. S. "Medical applications of wireless body area networks." *Int. J. Digital Content Technol. Applications*, 2009.
38. Chen, M., Gonzalez, S., Vasilakos, A., Cao, H., Leung, V. C. "Body area networks: A survey." *Mobile Networks and Applications* Vol. 16, No. 2, 171–193, 2011.
39. Md. Tanvir Ishtaiqueul Huque, Kumudu S. Munasingheand Abbas Jamalipour. "Body Node Coordinator Placement Algorithms for Wireless Body Area Networks." *IEEE Internet of Things Journal* Vol. 2, 94–102, 2015.
40. Binh, H. T. T., Hanh, N. T., Dey, N. "Improved cuckoo search and chaotic flower pollination optimization algorithm for maximizing area coverage in wireless sensor networks." *Neural Computing and Applications* 1–13, 2016.
41. A. Pantelopoulos, N. G. Bourbakis. "A Survey on wearable systems for monitoring and early diagnosis for the elderly." *IEEE Transactions on Systems, Man and Cybernetics, Part C: Applications and Reviews* Vol. 1, 1–12, 2010.
42. D. Dakopoulos, N. G. Bourbakis. "Wearable obstacle avoidance electronic travel aids for visually impaired: A Survey." *IEEE Trans on Systems, Man and Cybernetics* Vol. 40, 25–35, 2010.
43. S. Ha, E. Wyat, N. Javaid, Z. A. Khan, A. Shareef, A. Mahmood, S. H. Bouk. "Energy Efficient MAC Protocols." IEEE 14th International Conference on High Performance Computing and Communication, 2012.
44. Okundu Omeni, Alan Chi Wai Wong, Alison J. Burdett, Christofer Toumazou. "Energy Efficient Medium Access Protocol for Wireless Medical Body Area Sensor Networks." IEEE Transaction on Biomedical circuits and systems Vol. 2, No. 4, Dec 2008.
45. Nourchne Bradai N., Lamia Chaari Fourati, Lotfl Kamoun, "Investigation and performance analysis of MAC protocols for WBAN networks." *Elsevier* 1084–8045, 2014.
46. Bin Liu, Zhisheng Yan, Chang Wen Chen. "Mac Protocol in Wireless body area networks for E-Health: Challenges and a context-aware design." *IEEE Wireless Communication*, Aug 2013.
47. Samaneh Movassaghi, Mehran Abolhasan, Justin Lipman, David Smith, Abbas Jamalipour. "Wireless Body Area Networks: A Survey." *Communications Surveys & Tutorials*, IEEE, 16, 2014.

48. Samaneh Movassaghi, Mehran Abolhasan. "A review of routing protocol in wireless body area network." *Journal of Network* Vol. 8, 2013.
49. Javed Iqbal Bangash, Abdul Hanan Abdullah, Mohammad Hossein, Anisi, Abdul Waheed Khan. "A Survey of Routing Protocols in Wireless Body Sensor Networks." *Sensors* 1322–1357, 2014.
50. J. Su Kim, J. Hyun Lee, K. Wook Rim. "Energy Efficient Key Management Protocol in Wireless Sensor Networks." *IJSIA* Vol. 4, No. 2, Apr 2010.
51. A. Bag, M. A. Bassiouni. "Energy efficient thermal aware routing algorithms for embedded biomedical sensor networks", IEEE International Conference on Mobile Adhoc and Sensor Systems (MASS), 604–609, Oct 2006.
52. Wannenburg, J., Malekian, R. "Body sensor network for mobile health monitoring, a diagnosis and anticipating system." *IEEE Sens. J.* Vol. 15, 6839–6852, 2015.
53. R. Shah, M. Yarvis. "TARA: Thermal-aware routing algorithm for implanted sensor networks." 1st IEEE International Conference on Distributed Computing.
54. J. Sung Choi, J. Gon Kim. "An Improved MAC Protocol for WBAN through modified frame structure." *International Journal of Smart Home* (IJSH) Vol. 8, No. 2, 2014.
55. Nabi, M., Basten, T., Geilen, M., Blagojevic, M., Hendriks, T. "A robust protocol stack for multi-hop wireless body area networks with transmit power adaptation." *Proceedings of the Fifth International Conference on Body Area Networks*, ACM, 77–83, 2010.
56. O. G. Morchon, H. Baldus, D. S. Sanchez. "Resource Efficient Security for Medical Body Sensor Networks, Wearable and Implantable Body Sensor Networks." BSN International Workshop, 4 pp, Apr 2006.
57. S. A. Gopalan, J. Park. "Energy Efficient MAC Protocols for Wireless Body Area Network: Survey" International Congress on Ultra-Modern Telecommunications and Control Systems and Workshops, pp. 739–744, Oct 2010.
58. Javaid, N., Yaqoob, M., Khan, M. Y., Khan, M. A., Javaid, A., Khan, Z. A. "Analyzing delay in wireless multi-hop heterogeneous body area networks." *Research Journal of Applied Sciences, Engineering and Technology*, 2013.
59. Javaid, N., Abbas, Z., Farid, M.S., Khan, Z.A., Alrajeh, N. "M-ATTEMPT: A new energy-efficient routing protocol for wireless body area sensor networks." The 4th international conference, ANT, 224–231, 2013.
60. A. Pantelopoulos, N. G. Bourbakis. "A Survey on wearable systems for monitoring and early diagnosis for the elderly." *IEEE Transactions on Systems, Man and Cybernetics, Part C: Applications and Reviews* Vol. 1, 1–12, 2010.
61. D. Dakopoulos, N. G. Bourbakis. "Wearable obstacle avoidance electronic travel aids for visually impaired: A Survey." *IEEE Trans on Systems, Man and Cybernetics* Vol. 40, 25–35, 2010.
62. A. Saeed, et al. "A Scalable Wireless Body Area Network for Bio-Telemetry." *Journal of Information Processing Systems* Vol. 5, 77–86, Jun 2009.
63. Jin, Z., Oresko, J., Huang, S., Cheng, A. C. "HeartToGo: A personalized medicine technology for cardiovascular disease prevention and detection." *Life Science Systems and Applications Workshop*, IEEE 80–83, Apr 2009.
64. Hadjidj, A., Souil, M., Bouabdallah, A., Challal, Y., Owen, H. "Wireless sensor networks for rehabilitation applications: Challenges and opportunities." *Journal of Network and Computer Applications* Vol. 36, No. 1, 1–15, 2013.

65. Enrique Gonzalez, Raul Pena, Cesar Vargas-Rosales, Alfonso Avila, David Perez-Diaz de Cerio. "Survey of WBSNs for Pre-Hospital Assistance: Trends to Maximize the Network Lifetime and Video Transmission Techniques." *Sensor MDPI*, 2015.
66. Elhayatmy, G., Dey, N., Ashour, A. S. "Internet of Things Based Wireless Body Area Network in Healthcare." *Internet of Things and Big Data Analytics Toward Next-Generation Intelligence*. 3–20. Springer, Cham, 2018.
67. Yadav, P., Sharma, S., Tiwari, P., Dey, N., Ashour, A. S., Nguyen, G. N. "A Modified Hybrid Structure for Next Generation Super High Speed Communication Using TDLTE and Wi-Max." *Internet of Things and Big Data Analytics Toward Next-Generation Intelligence*. 525–549. Springer, Cham., 2018.
68. Redondi, A., Tagliasacchi, M., Cesana, M., Borsani, L., Tarrío, P., Salice, F. "LAURA—LocAlization and Ubiquitous monitoRing of pAtients for health care support." *Proceedings of the 2010 IEEE 21st International Symposium on Personal, Indoor and Mobile Radio Communications Workshops (PIMRC Workshops)*, Istanbul, Turkey, 26–30, pp. 218–222, Sep 2010.
69. Jachimczyk, B., Dziak, D., Kulesza, W.J. "RFID-hybrid scene analysis-neural network system for 3D indoor positioning optimal system arrangement approach." *Proceedings of the 2014 IEEE International Instrumentation and Measurement Technology Conference* (I2MTC), Montevideo, Uruguay, 12–15, pp. 191–196, May 2014.
70. Kang, W., Han, Y. "SmartPDR: Smartphone-based pedestrian dead reckoning for indoor localization." IEEE Sens. J. 15, 2906–2916, 2015.
71. Shchekotov, M. "Indoor localization methods based onWi-Fi lateration and signal strength data collection." *Proceedings of the 2015 17th Conference of Open Innovations Association* (FRUCT), Yaroslavl, Russia 20–24, pp. 186–191, Apr 2015.
72. Palumbo, F., Barsocchi, P., Chessa, S., Augusto, J. C. "A stigmergic approach to indoor localization using Bluetooth Low Energy beacons." *Proceedings of the 2015 12th IEEE International Conference on Advanced Video and Signal Based Surveillance* (AVSS), Karlsruhe, Germany, 25–28, pp. 1–6, Aug 2015.
73. Wyffels, J., Brabanter, J. D., Crombez, P., Verhoeve, P., Nauwelaers, B., Strycker, L. D. "Distributed, signal strength-based indoor localization algorithm for use in healthcare environments." *IEEE J. Biomed. Health Inform.* 18, 1887–1893, 2014.
74. Shen, X., Chen, Y., Zhang, J., Wang, L., Dai, G., He, T. "BarFi: Barometer-aidedWi-Fi floor localization using crowd sourcing." *Proceedings of the 2015 IEEE 12th International Conference on Mobile Ad Hoc and Sensor Systems* (MASS), Dallas, TX, USA, 19–22, pp. 416–424, Oct 2015.

Index

L

M

N

O